NUMBERS

NUMBERS

Henry F. De Francesco

Copyright © 2015 by Henry F. De Francesco.

Library of Congress Control Number: 2014921409
ISBN: Hardcover   978-1-5035-2149-0
            Softcover   978-1-5035-2150-6
            eBook   978-1-5035-2151-3

All rights reserved. No part of this book may be reproduced or transmitted in any form or by any means, electronic or mechanical, including photocopying, recording, or by any information storage and retrieval system, without permission in writing from the copyright owner.

Print information available on the last page.

Rev. date: 03/20/2015

To order additional copies of this book, contact:
Xlibris
1-888-795-4274
www.Xlibris.com
Orders@Xlibris.com
697725

Henry F. De Francesco
henbob2@aol.com

CONTENTS

List of Figures 11

Acknowledgments 13

Introduction 15

Preface 23

Numbers 27
    Introduction 27

Sets 31
    Set Definitions 33

Set Operations 35
    Complement 35
    Intersection of Sets 35
    Disjoint Sets 36
    Union of Sets 36
    Difference 37

Elements of Logic 39
    Statements 39
    Truth Tables 41
    Conditional 46
    Valid Arguments 47

Plausible Reasoning 49
    Introduction 49
    Report Credibility 50
    Information Synthesis 52

    Information Credibility 52
    Use of Kent Chart 54
    Interchangeability 57
    Credibility Scale 58
    Relational Evaluation 59
    Credibility Patterns 61
    Credibility Relationships 63
    Deductive Patterns 65
    Plausibility Synthesis 67

Numbers 71
    Executive Summary 71

Classes of Numbers 73
    Zero 73
    Natural Numbers (N) 74
    The Peano Axioms 75
    Integers (Z) 77
    Decimal Numbers (D) 77
    Rational Numbers (F) 78
    Algebraic Numbers (A) 79
    Real Numbers (R) 80
    Irrational (I) and Transcendental Numbers (T) 83
    Prime Numbers 86

Number Systems 89
    Decimal Numbers 89
    Binary Numbers 92
    Converting Base 2 Numbers to Base 10 94
    Converting Base 10 Numbers to Base 2 95
    Highest Power Division Algorithm 96
    Division by 2 Algorithm 97

Converting Decimal Numbers 99
    Highest Power Division Algorithm 99
    Division by 2 Algorithm 101

Cardinal Numbers 102
    Power Set 104
    Exponentiation 106

Appendix 1 109
    1. Converting Cyclical Decimals to Rational Form (p/q) 109
    2. Rational Numbers Have Cardinal Number N_0 111
    3. Algebraic Numbers 115
    4. Square Root of 2 118
    5. Irrational and Transcendental Numbers 120
    6. Line and Plane Have Same Cardinal Number 123

In memory of my wife,
Bobbie J. Neal De Francesco,
June 6, 1926–December 19, 2012

LIST OF FIGURES

Body of text

Figure 1. Shepherd tallying sheep 28
Figure 2. Real number continuum 32
Figure 3. Complement..35
Figure 4. Intersection 36
Figure 5. Disjoint 36
Figure 6. Union..37
Figure 7. Difference 37
Figure 8. Statement development tree..40
Figure 9. Composite truth table..42
Figure 10. Implication truth table...43
Figure 11. Equivalent statements 44
Figure 12. Associated statements 45
Figure 13. Biconditional statements.. 46
Figure 14. Valid arguments..48
Figure 15. Credibility code 51
Figure 16. Modified Kent chart...53
Figure 17. Combined charts...56
Figure 18. Interchangeability...58
Figure 19. Credibility scale 59
Figure 20. Paired credibility scales..59
Figure 21. Credibility relationships..60
Figure 22. Credibility relationships 64
Figure 23. Credibility regions 65
Figure 24. Deductive patterns 66
Figure 25. Subsets of R with cardinal number N_0 81
Figure 26. Cantor's diagonal process 83
Figure 27. Diagonal of square..84
Figure 28. Subsets of R with cardinal number N_0 and c............85
Figure 29. Eratosthenes's sieve 87

Appendices

Figure A2.1. Triangular numbers 112
Figure A2.2. Plot of rational numbers 113
Figure A5.1. Cantor's diagonal process 120
Figure A6.1. Line and square 123
Figure A6.2. Line and plane have same cardinal number. 124

ACKNOWLEDGMENTS

I WISH TO thank John Wiley and Sons Inc. for permission to reproduce parts of the book titled *Quantitative Analysis Methods for Substantive Analysts* by Henry F. De Francesco, ISBN 0 471-20529-X

I wish to thank Perry Milou for his permission to use his unique sketch for figure 1 of the monograph.

I wish to thank my friend and neighbor, Barbara Sheffield Smith, for the time and talent she devoted to the design of the monograph's beautiful cover.

I wish to thank members of the Xlibris staff for professionally guiding me through the entire preparation and production process.

Finally, I wish to express my deepest appreciation and thanks to my daughter, Robbye M. De Francesco, for her patience and untiring efforts and talents in preparing the illustrations and editing the many drafts of this monograph.

INTRODUCTION

Why This Monograph?

THE SAGA LEADING up to the writing of this monograph started when the President's Foreign Intelligence Advisory Board (PFIAB) recommended that the Intelligence Community initiate a training program in the information sciences. I was hired by the Defense Intelligence Agency (DIA) as director of the yet-to-be-formed Information Science Center (ISC). The ISC provided training in the information sciences for the entire Intelligence Community. The ISC student population included both military and civilian intelligence analysts. Almost all the analysts had a bachelor or higher degree in a major other than the sciences. Most were considered experts in their field of study. Over the years they had collected and compiled reams of intelligence on most countries and international persons of interest. Each analyst maintained their store of information in personal Rolodexes, boxes, and filing cabinets. The ISC had to train the intelligence analysts how to evaluate, store, and make their information more readily accessible to other *need-to-know* analysts.

Most of the courses offered by the ISC were on the utilization of computers to store, process, and disseminate information. The course I prepared and offered dealt with how to utilize sets, logic, and plausible reasoning in the evaluation and processing of information. There were several analysts in the class that had an extreme dislike or aversion to mathematics. At the end of the course, these analysts confessed to having profited from the course. The same result was obtained from a second offering of the course. It was the analysts' acceptance of the course content that led me to write my first textbook. However, that book also emphasized statistics and probability, subjects still considered mathematics by those having an aversion to math. Going through the old text, I decided to write a new monograph based solely on a less technical version of the first three chapters of the old text. Axiom systems, sets, and logic form part of the foundations for mathematics. I wanted to show how the foundations of mathematics are used to develop fields of mathematics. I selected the real numbers since virtually everyone is familiar with some of the elements of the real number continuum. Also, it was easy to show how axioms, sets, and logic are used in defining and constructing elements of the real number system.

Did I succeed in meeting my objectives? I leave it to the readers to provide answers to the question.

We Are All Analysts

As intelligent individuals, we consider ourselves capable of analyzing problems when they arise. We think we have the necessary analytical tools to make informed decisions. We make decisions based on our analysis of information provided by a variety of sources. The sources include friends, family members, coworkers, news items, advertisements, research articles, politicians, professional advisors, and many others. Each source prepares and presents information using one or a combination of four types of analysis. Briefly, the four types of analysis are *descriptive, predictive, normative, and prescriptive.* When analyzing information sources, are we able to identify which of the four types of analysis was used by the source? For most individuals, the answer is no. Descriptions of the four types of analysis are provided below.

Do you consider yourself an effective analyst? As an analyst, you must seek out and examine all available information. As an analyst, you have to be concerned with questions of substantive content and with analytic processes performed on this content. Based on the substantive content of the information, you have to make assessments and predictions. Then you must effectively communicate your findings to interested parties. For example, as analysts we all have to deal with substantive information from the political, social, military, diplomatic, geographical, historical, and business communities. The information will, at times, contain incomplete and sometimes corrupted information. We have to be able to distinguish fact from fiction. As innocent consumers, we are faced with the same problems when attempting to purchase a new automobile or deciding how to vote in an election.

By employing elements of the deductive, inductive, and plausible reasoning processes, as recommended in the monograph, you, as an analyst, will have the analytical tools to effectively assess substantive information. Good information poorly analyzed leaves much to be desired. But poor information properly analyzed and evaluated can contribute to the understanding of the substantive content and its implications, even if only to determine the areas wherein more information is required. The sections of the monograph on sets, logic, and plausible reasoning provide analysts with some of the basic methods and techniques that have been proven useful in the analysis of subjectively and objectively derived information.

Information properly analyzed and disseminated will achieve one or a combination of the following objectives:

1. Enhanced understanding of the past and present
2. More accurate predictions about the future
3. Identifying common ground among different points of view
4. Establishing common goals and prescribing plans through which these goals can be achieved

The four types of analysis, descriptive, predictive, normative, and prescriptive, are described below.

Descriptive Analysis

The type of analysis most frequently used by analysts is termed *descriptive*. The objective is to describe the characteristics of things, statements, or events on the basis of historical or current observations. A description usually consists of two components: one describes the event, statement, or thing; the other rationalizes the existence of the thing, the truthhood of the statement, or the occurrence of the event.

Descriptive analysis uses a spectrum of individualized styles and a modicum of techniques and methods. For example, intuition and heuristics (educated guesses) play an important role in the development of ideas that lead to more accurate descriptions. Several pieces of information may be tied together deductively or inductively, yielding an input to a better description of the thing, statement, or event. In the worst case, reams of information will be studied in an attempt to make some useful generalizations about the properties or characteristics of the thing, statements, or events under observation.

Descriptive analysis normally takes the form of a verbal description, a statistical description, or a combination of the two. Non-scientifically-oriented analysts use verbal descriptive analysis most frequently in their work. Training in this technique is acquired mainly through experience or courses in writing and communicating. This aspect of descriptive analysis will always be an important output of any analyst attempting to communicate his values, thoughts, or findings to the interested but not professionally prepared public. The scientific analyst would do well in learning the techniques of good verbal descriptive analysis.

Predictive Analysis

The analyst must of necessity be anticipatory or predictive of things developing or events occurring. For example, to answer questions such as "How will the Dow-Jones average react to an increase in interest rates?" or "If Congress passes an immigration reform act, what will be the effect on the political parties chances in the next election?" the analyst will employ what is called *predictive* analysis. The analyst normally attempts to describe the thing or event as it might appear to exist in the future.

Descriptions of things or events as they existed in the past and as they currently exist and the rate of change in key properties from the past to the present guide the analyst in deciding whether the same rate of change (trend) will continue into the future. For example, an analyst has information on the period from 1950 to 2000. He uses that information to set up and check the validity of a trend occurring in the period 1950–2000. He uses that trend to predict what will happen in the period 2000–2010. His descriptive analysis covers the period 1950–2000. On the basis of that description and the predictive analysis of that information, he can make assertions about the period 2000–2010. If his description and methods yield predictions about the period 1950–2000 that are credible (even if less than completely accurate) in that they predict the events that actually occurred in the 1950–2000 period, then his confidence in making predictions out to the year 2010 is enhanced proportionately.

Normative Analysis

Another type of analysis used by almost everyone and, hopefully, to a lesser degree by trained analysts is called *normative* analysis. It is analysis based on one's own value system. In normative analysis, judgments are formed from one's personal view of what exists or what will exist. Normative analysis reflects in varying degrees personal preferences, be they religious, legal, moral, political, or social. Normative analysis reflects an individual's view of reality tempered by his notions of right and wrong, good and bad, or acceptable and unacceptable. These notions become the main theme of his or her reasoning processes.

For example, someone who believes strongly that communism is the greatest evil confronting the free world would have a very difficult time objectively evaluating any benevolent act by a communist nation. Most descriptions of the act and reasons for the act would imply, if not directly state, that the acting nation had some devious or ulterior motive other than the one of pure benevolence. No individual is void of normative analytic techniques in employing reasoning processes. Each of us, in our growing-up process, has formed sets of hard-core values. How these values are acquired is of utmost importance. Briefly then, as stated in the section on logic, if the hypothesis (our value system) does not reflect reality, then even the best logical thought processes employing this value system cannot guarantee a rational, let alone a true, description or prediction.

Prescriptive Analysis

Another type of analysis is called *prescriptive*. Prescriptive analysis is used to demonstrate how and why future goals, policies, values, and strategies should assume the forms that the analyst considers to be desirable. Normally, prescriptive analysis includes a suggested plan of action on how to accomplish the analyst's desired objectives. Descriptive, normative, and predictive analytic components are also present in the plan, and analysis to indicate why these forms are desirable and should be sought. There are several techniques used in prescriptive analysis. The techniques include implication, suggestion, and generalization. Their use is illustrated by means of the following example.

The United States, at the end of World War II, set for itself the following aims:

Goal: Containment of communism to existing boundaries
Value: Self-determination and individual freedom
Policy: Support and development of needy noncommunist nations
Strategy: World peace through alliances of "free world" nations

Initially, the United States used the implication form of prescriptive analysis to illustrate how foreign aid would serve the United States in achieving the goal, value, policy, and strategy stated above. The foreign aid program was sold to the public (and Congress) solely on the implication that if the stated aims of our nation were achieved, then the world would be a better place in which to live. The United States decided that the quickest way to achieve these aims was through an extensive foreign aid program. In other words, the United States prescribed a better world and implied that its plan for foreign aid was the best one to achieve this major aim.

After foreign aid had been in effect for several years, its continuance to nations not cooperating fully was used as a means to suggest to these nations the desirability of undertaking actions favorable to the goals of the United States. This constituted the suggestive form of prescriptive analysis.

After foreign aid had been in effect for a considerable period of time, the United States felt itself burdened most heavily and made its general prescriptive analysis, which stated that if the free world were to achieve the previously stated four aims, all free nations should share in the aid to other free nations requiring assistance.

Responsibility of the Analyst

Every analyst should be familiar with each of these analysis forms and their special attributes. Further, the analyst should be aware of the type of analysis he is performing. To mix or confuse forms of analyses results in a misunderstanding of one's stated position, predictions, values, and prescriptions. It is most important that the analyst understand whether the statements are descriptive, predictive, normative, or prescriptive. Through this knowledge, the analyst achieves greater objectivity in his or her analysis and impresses others by being able to identify their method of analysis.

After reading about the four types of analysis you may be wondering why I titled the monograph NUMBERS. In a logical or mathematical context, sets and terms are clearly defined and statements are either true or false. In an everyday context words have a range of meanings and statements have a range of credibility. Plausible reasoning identifies how to assign values other than true or false to statements. It also establishes a pseudo- logical structure to analyze composite statements. In both cases the assigned values are numerical replacing standard nominal values.

The section of the monograph titled Plausible Reasoning (pages 49-67) introduces the Credibility code and the Kent chart. When combined they show how the results of any analysis, normally expressed in nominal terms, can also be expressed in numerical terms. It is generally accepted that results or conclusions expressed in numerical terms convey more information, and hence are accepted with more confidence, than the same conclusions expressed in nominal terms.

While the introduction emphasizes the importance of knowing the four types of analysis, the main focus of this monograph, as illustrated in the section on plausible reasoning, emphasizes the importance of using numerical values in explaining the results of your analysis. The use of the four types of analysis is old hat. The use of numbers, i.e., numerical values, in place of nominal values is new to most of the readers of this monograph. Hence the title NUMBERS.

PREFACE

YOU DON'T HAVE to be an expert in trigonometry, geometry, or calculus to understand the content of this monograph. A little simple algebra will do. As you delve into this monograph, your vocabulary will expand to include many new terms not normally encountered in the standard courses of mathematics. This monograph should be viewed as an introduction to a language of sets and logic—a language used to explain how and why math works.

In preparing this monograph, I had three objectives. First, I wanted to introduce the reader to some topics in mathematics that are seldom covered in typical high school and college math programs. The topics include axioms, sets, logic, truth tables, and plausible reasoning. In the sections on logic and plausible reasoning, I wanted the reader to see how to transition from formal (mathematical) logic to plausible logic when analyzing the reliability of a source and the credibility of its information content. Readers whose formal education did not cover these topics were not given the opportunity to develop the skills necessary to compete successfully in the world of finance, business, and management. These readers will find the information on sets, logic, truth tables, and plausible reasoning especially useful. Included are examples that show how the new analysis skills can help analysts draw conclusions and make important decisions from subjective information supplied by less than reliable sources.

Second, I wanted the reader to see how subjects in the foundational area of mathematics are used to develop the real number system and its extension through transfinite cardinal numbers. The development of the number system starts with a description of the history of numbers. Readers will find the history both interesting and understandable. The real number continuum is identified as consisting of seven sets of numbers. Each set of numbers can stand alone. The number sets include the simple to understand natural numbers to the more abstract transcendental numbers. Each set is defined and included in a vocabulary consisting of the natural numbers (*N*), integers (*Z*), the rational numbers (*F*), the algebraic numbers (*A*), transcendental numbers (*T*), irrational numbers (*I*), and real numbers (*R*). Venn diagrams are used to explain the relationships existing among the seven sets. The relationships allow the reader to understand the role played by sets and logic in the development of the number system. Included in the development of the real number system are examples of base 2 numbers and the algorithms used to convert between base 2 and base 10 numbers. Power sets are introduced to show how the size of sets can be increased exponentially beyond the cardinal numbers N_0 and c. Finally, through exponentiation, cardinal numbers are generated beyond the $N_0 < c < f$ sequence.

Third, I wanted this monograph to appeal to those adults, and their children, who have an antimath bias. This bias is exhibited as innumeracy or an aversion to math. In either case, those afflicted find it difficult to compete against the mathematically literate in the world of business, finance, and technology. Through this monograph, I attempt to address this antimath bias. The reader is introduced to the language of sets, logic, and plausible reasoning. While these subjects are part of the foundations of mathematics, they are also subjects taught in philosophy departments without math prerequisites. The reader is then shown how axioms, sets, and logic are applied in the development of the number system. The subject of numbers is made intelligible for a broader spectrum of readers through the use of verbal descriptions and graphics, rather than equations, wherever possible. To achieve a continuous flow of understandable subject matter, the more tenuous procedures and methods of mathematics are explained in six appendices. Through understanding how math works at the fundamental level and not having to work at math, the intelligent reader's antimath bias should be reduced, if not eliminated.

This monograph is a nonresearch type of document. I consider references placed in nonresearch type-documents as distractions and interruptions to learning and thought continuity. As a result, this monograph does not list any references. I had the choice of placing my references at the back of the text, but that list would be limited to my choices. I want to acknowledge that a better and more complete list of references can be found on the Web—that wonderful electronic and digital universe of knowledge. If you would like to delve further into any subject covered in this monograph, I recommend that you search the Internet using entries in the Contents section as key search words or phrases. I guarantee that you will uncover information that will answer any question you may have on the subjects covered in this monograph. For more definitive books on logic, plausible reasoning, and foundations of mathematics, I recommend Quine, W. V. O.; Polya, G.; and Wilder, R. L.

Who should read this monograph?

- Any professional decision maker who would like to learn new analysis tools that can assist him or her in evaluating subjective information from unreliable sources.
- Anyone who would like to see how the foundations of mathematics may serve as a basis from which he or she may improve his or her writing and reasoning skills.
- Any teacher of math who would like to see how math skills can be taught using other than the normal dogmatic approach.
- Anyone with an antimath bias who would like to see how the foundations of mathematics are applied in developing the number system and reasoning skills without the use of advanced algebra and calculus.

NUMBERS

Introduction

LEOPOLD KRONECKER—A FAMOUS German number theorist and mathematician who lived from December 7, 1823, to December 29, 1891—is credited for saying "God made the natural numbers, everything else is the work of man." A simple interpretation: "The natural numbers are abstractions of nature." The mysterious number concept evolved when man gained possession of many objects. Record keeping and trading of possessions required that amounts be noted. Numbers were used without knowledge of their logical basis. Along with the development of learned societies, numbers assumed a greater utilitarian role as well as a philosophical and religious role. Historians have noted that the Babylonians developed a special number system for their studies of astrology. The Egyptians developed number concepts for their use in measurements. The Greek Pythagoreans contained some religious cults that worshipped numbers. However, the number 0 baffled them. They questioned, "How could nothing have any meaning?"

From the earliest civilizations to about 3400 BC (Mesopotamia), numbers were mainly tallying systems used to keep track of time (days) and possessions (livestock, wealth). The following is an example of a tally system that could have been used by shepherds keeping track of their flocks: The tally took place when the sheep left an enclosure and returned after feeding in the field. The shepherd assembled a cluster of pebbles equal in number to his sheep count. He selected two locations to store his pebbles. As each sheep left the enclosure, the shepherd moved a pebble from one location to the other. The process continued until all sheep left the enclosure. At the end of day, as each sheep was returned to the enclosure, the shepherd returned a pebble to its original position. When all pebbles were returned to their original pile, the shepherd knew all sheep had returned to the enclosure. If pebbles remained, the shepherd knew he had to go out with his sheepdogs and find the wayward sheep.

This tallying process is called a one-to-one correspondence. This process of tallying one set of items with a different set of items works OK when keeping track of a small number of items. Other methods are necessary when a large number of items are to be tallied. A wise shepherd uses his fingers to tally up to ten sheep. He then employs a cohort to raise a finger each time he completes the ten-finger count. This process goes on until the cohort has raised all his fingers, at which time the tally is one hundred. A second cohort is instructed to raise a finger each time the first cohort raised all his fingers. In this manner, the shepherd keeps track of up to a one thousand sheep. With the use of each additional cohort, the number of sheep that could be tallied increases by a factor of ten. Note the importance of each shepherd's position in completing the tally.

Figure 1. Shepherd tallying sheep

In the simple direct tally system (pebbles for sheep), there is no counting as we now know it. A completed pile of pebbles serves as the graphic for the final head count. What has taken place is a pairing of pebbles with sheep. When the fingers of several shepherds are used to tally the sheep, the boss shepherd has to use a unique notational system (graphic) to remember the finger positions at the final tally count.

In this example, the graphic would represent a three-finger, no-finger, and five-finger configuration. Using our current decimal notation, the shepherd has tallied 305 sheep. He did not count to 305. He had no knowledge of the digits 0 to 9 and their use in a counting process. A merchant, in the same enclave as the shepherd, could use a different tallying system to package corn to be sold in the market place. If he tallies 305 ears of corn, there is no assurance that his notation (graphic) for 305 would be the same as the shepherd's. Both used a one-to-one correspondence to get to the final total. The number concept, as we know it, does not enter into the matching. The number concept evolves when the shepherd, the merchant, and others agree to use the same graphics to represent their final totals, regardless of where the tallying stops. Each graphic in the system of graphics represents the same tally for whatever objects are being tallied. We now call that graphic a number. It identifies the property that a group of items has with every other group of items with which it can be placed in a one-to-one correspondence. Today, that number property has a special name. It is called a cardinal number. The natural numbers are cardinal numbers (1, 2, 3). The cardinal numbers represent the property possessed by all groups whose items can be placed in a one-to-one correspondence. We normally call the common property of the groups their size.

In our elementary educational system, numbers are introduced and taught with little or no explanation. We learn both numbers and the alphabet by rote. Classrooms for beginning students have numbers and letters prominently displayed on their walls. We are taught how to memorize numbers and letters, then we are taught how to use them. With numbers, we first learn simple counting, then we use numbers as a measure to determine sizes. With arithmetic tables, we learn how to compute amounts. We accept that we can count from the number 1 forever to that unattainable limit we call infinity. We never did learn much about infinity. Our introduction to infinity was the teacher's admonishment, "Never divide by 0." An objective of this monograph will be to correct that shortcoming by explaining, in simple terms, the number continuum from 0 to cardinal numbers and thence to transfinite numbers. However, division by 0 is still verboten.

To accomplish the objective of this monograph, you will be introduced to Venn diagrams and their use in explaining the most basic and simple elements of set theory. With only a basic introduction to sets, you should be able to understand how the real number system can be broken into at least seven sets. Once broken into sets, you will understand how these sets are related and can be ordered as to size. You will learn that numbers are used to identify a property of sets of objects. For example, the number 6 identifies a set of six objects. The set may consist of one type of item or a mixture of types. Knowing the content type in the set is not relevant to the number assignment. The property of numbers to be emphasized is that a number is a graphic. The graphic identifies the property held by sets of elements. Sets are assigned the same graphic (number) when their elements can be placed in a one-to-one correspondence, i.e., when it is possible to pair the elements of the sets element by element. The process is obvious for sets having a finite number of elements. It is more difficult, but still possible to do, with sets that are not finite. You will learn that the natural numbers (1, 2, 3,…) are an infinite set and serve as a basis to define sets having higher order infinities. To guide our journey into the set of higher order infinities, we need to review some basic properties of sets.

SETS

YOU MAY ASK, "What does the theory of sets have to do with understanding the structure of the number system?" The answer is simple. Set theory is an integral part of the foundations for mathematics. The various elements of the number system are best described using the set notation. The theory of sets establishes the rules used to determine whether an element qualifies for membership in a set. In common usage, the term "set" is used loosely and synonymously with collection, class, aggregate, and family. However, formal usage of the term "set" requires that the elements of the set be accurately defined and partitioned according to predetermined properties. The terms "define" and "partition" are the properties of sets that are not only fundamental to math and science but also to everyday activities. The more you know how to use set concepts, the better you are at explaining your thoughts and organizing your life.

Simple functions such as organizing your closets, searching the Internet for information, and describing your automobile to others are best performed using the rules for set formation. Even your home and furniture use set principles in their design. Each room in your home was designed with a specific purpose in mind. The triple dresser in your bedroom was designed with nine drawers to allow you to separate (partition) your personal belongings, if you want to, into nine sets. Jewelry in one drawer, socks in another, and shirts or blouses in another. Each item is uniquely defined and partitioned according to definition. However, you are not required to segregate your items into distinctly designated sets. You can mix or match the items you place in each drawer. That is not the case in mathematics. In math, precision is paramount in defining sets. Sets are defined by describing the properties an element must possess in order to qualify for membership in the set.

For the purposes of this monograph, you do not have to be an expert in set theory. You can restrict your introduction to set theory to those set concepts necessary to understand the partitioning of the real number system into its subclasses. When dealing with collections of subsets, it is advisable that the collection be considered as belonging to a larger set. For example, the natural numbers (*N*) are considered a subset of the real numbers (*R*). When a set of elements is restricted in membership to a larger set, the larger set of element, to which discourse is limited, is called the universal set. Unless otherwise stated, the set of real numbers (*R*) will serve as our universal set for the subsets identified below.

Some of the subsets of the universal set of real numbers are listed below:

1. The natural numbers 1, 2, 3 are designated *N*.
2. The integers –3, –2, –1, 0, 1, 2, 3 are designated *Z*.
3. The rational numbers are ratios of two integers (p/q, q not 0) and are designated *F*.
4. The algebraic numbers are roots of the general polynomial equation, designated *A*.
5. The transcendental numbers are real numbers that are not algebraic numbers, designated *T*.
6. The irrational numbers are real numbers that are not rational numbers, designated *I*.
7. The prime numbers are natural numbers that are not factorable, designated *P*.
8. The null set is the set having no elements (designated 0).

A unique element of any universal set is the null or empty set, denoted 0, the set having no elements. The null set is a subset of every set. However, the null set is not an element of every set.

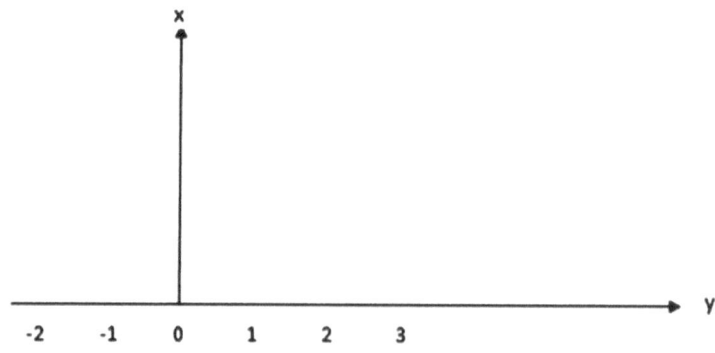

Figure 2. Real number continuum

We assume that you are familiar with the real numbers (R). As shown in figure 2, they are normally identified as the names for points on the x-axis of the x-y coordinate system. This axis, or the interval between any two consecutive integers on it, is normally called the real number continuum. In the sequel, we will deal with natural numbers (N) as a subset of real numbers (R). The natural numbers (N) will serve as a basis for defining finite and infinite sets. To simplify our explanations of the operations that can be performed on and between sets, we will restrict our discussions to the real numbers (R) and their subsets.

Set Definitions

One may define a set as a collection of definable objects. The objects may be real or contrived but must have definable properties. To operate with sets, it is necessary to have a notation to designate the set and its contents. In general, the following notation is used to designate sets:

$$S = \{x \mid x \text{ has property } (P)\}$$

The notation reads as follows: the parentheses {} denotes a set, the x denotes elements of the set, the vertical bar reads "such that," and what follows identifies the properties the elements (x) must have to qualify as members of the set (S). Briefly, the set (S) contains all elements (x) such that x has property (P). The properties (P) may be (1) tabulations or listings by names (e.g., numbers 1 to 13, etc.) or (2) by description or rules (e.g., odd numbers, prime numbers). To designate that an element (x) is a member of S, we use the symbol $\in$.

$$x \in S$$

For example, from the set of natural numbers (N), we can ask for the set.

$$S = \{x \mid x \in N, x \text{ is even}\}$$

The set (S) contains the natural numbers 2, 4, 6, ...

We accept the notion of a set as an intuitive term. As with point and line in plane geometry, the set concept will take on new meanings as you gain experience in their use and applications.

To designate that a set B is a subset of S—i.e., B is contained in S—we use the symbol $\subset$.

$$B \subset S \text{ or } S \supset B$$

The notation states that B is a subset of S if every element of B is an element of S. The set B can be equal to the set S, denoted $B = S$; i.e., every element of B is an element of S and conversely. If S has elements not in B, then the set B is called a proper subset of S. A vertical bar through the symbols $\notin$ and $\not\subset$ indicate, respectively, not an element of S ($x \notin S$) and not a subset of S ($B \not\subset S$). In the sequel, use will be made of Venn diagrams to display graphically the universal set and the five operations on sets identified in the following paragraphs. In a Venn diagram, a rectangle and its interior represent the elements of a universal set R. The subsets of R are represented by sets of distinct points or areas interior to the rectangle.

SET OPERATIONS

Complement

FIRST, WITH RESPECT to a universal set R of elements containing a proper subset S, the complement $\overline{S}$ of the set S, denoted by a bar over the letter, is defined as those elements of R not in S. In set notation,

$$\overline{S} = \{x \mid x \in R \text{ and } x \notin S\}.$$

Simply, the complement $\overline{S}$ of a set S contains all elements of R not in S. For example, relative to the set of natural numbers (N), the odd numbers are the complement to the set of even numbers.

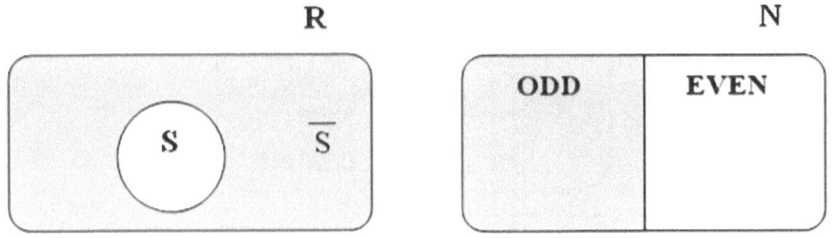

Figure 3. Complement

Intersection of Sets

The intersection of sets A, B, and C—denoted $A \cap B \cap C \cap$—is defined as the set of elements common to all the sets. In set notation,

$$A \cap B \cap C..., = \{x \mid x \in A \text{ and } x \in B \text{ and } x \in C \text{ and } ...,\}.$$

For example, if $A = \{2, 3, 5, 7, 9\}$; $B = \{3, 7, 9\}$; and $C = \{2, 5, 9\}$, the intersection $A \cap B \cap C = \{9\}$.

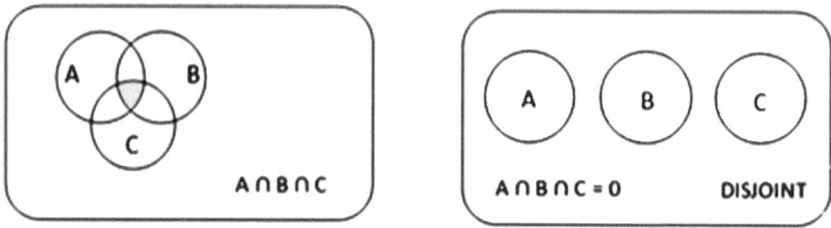

Figure 4. Intersection

Disjoint Sets

Two sets A and B are said to be disjoint if they have no elements in common. When their intersection $A \cap B = 0$, A and B are said to be disjoint. Disjoint sets are said to be mutually exclusive.

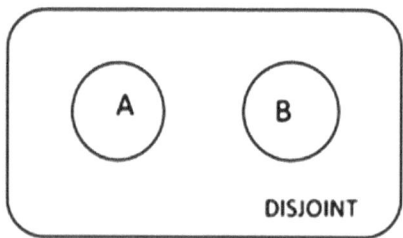

Figure 5. Disjoint

Union of Sets

The union (addition) U of sets A, B, C—designated $A \cup B \cup C$—is the set containing all elements that are members of at least one of the sets. In set notation,

$$U = \{x \mid x \in A \text{ or } B \text{ or } C \text{ or}...\}.$$

For example, the natural numbers (N) are the union of two subsets: the odd numbers (O), and the even numbers (E), where $N = O \cup E$.

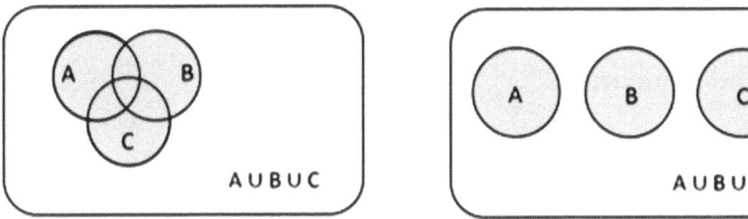

Figure 6. Union

Difference

The difference of two sets A and B, denoted $A - B$, is the set of elements belonging to A but not to B. In set notation, $A - B = \{x \mid x \in A \text{ and } x \notin B\}$. For example, the difference of the natural numbers (N) and the set of even numbers (E) is the set of odd numbers, $O = N - E$.

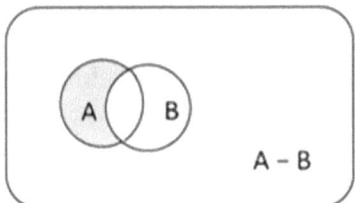
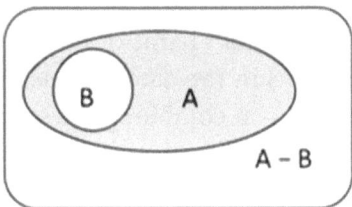

Figure 7. Difference

In the text, the first set we deal with is

$$N = \{x \mid x \text{ is a natural number}\}, N = \{1, 2, 3...\}.$$

We deal with subsets of N. A set B is a subset of N if the elements of B are elements of N. The notation for subset is $B \subset N$. Now B can be the same set as N. In this case, we state $B = N$. If not, then N has elements that are not elements of *B*. *B* is then a proper subset of *N*. Some proper subsets of *N* are the even numbers,

$$E = \{x \mid x = 2n, \text{ where } n \text{ is a natural number}\};$$

the odd numbers,

$$O = \{x \mid x = 2n - 1, \text{ where } n \text{ a natural number}\};$$

and prime numbers,

$$P = \{x \mid x \text{ has no factors other than 2}\}.$$

Operations with sets include addition, difference, intersection, negation, and complement. We have discussed these operations in prior sections.

The applicability of sets to daily life activities is illustrated through their use in searching for information in a library or on the Internet. When you attempt to search a library or Google the Internet for information, you will be using some of the most fundamental concepts of set theory. To accomplish a search, you identify the items you seek by using single or multiple terms, such as a product's name or manufacturer or an author's name or book title. The search results in a set containing all items in the database identified by the search terms. To refine your search, you combine your search terms by using set operations (logical connectives), such as union ("or"), intersection ("and"), and complement ("not"). For example, if *A* and *B* are items of interest, your search will ask the following types of questions: do you have items (authors, words, titles, or products) *A* or *B*? *A* and *B*? or *A* and not *B*? Your search will result in a set of replies containing more or fewer elements depending on how you use the logical connectives "or," "and," and "not."

ELEMENTS OF LOGIC

IN THE DISCUSSION on sets, we used the following notation,

$$S = \{x \mid x \text{ has property } (P)\},$$

to determine whether the element x qualifies as a member of the set S. The property (P) was described by statements or by rules. The statements could be simple or complex. Complex statements could be formed from simple statements using a combination of the connectives "and," "or," "not," and "difference." The resulting statements had to be unambiguous (true or false) to be of any use. Logic deals with the theory of statement composition. It is concerned with logical structures, which emerge when compound statements are formed from simple statements using basic locutions (particles) such as "and," "or," "not," "if-then," and "unless." Logic studies how logical structures, which are used to form compound statements from simple statements, determine the truth and validity of the compound statement based on the truth of its simple statements and the validity of the argument.

Statements

Statements (also called propositions) are declarative sentences to which a truth value can be assigned. A simple statement expresses a single fact. A compound statement expresses more than a single fact. The truth value of a statement is either truth or falsehood according to whether the statement is true or false. Figure 3 illustrates the relationships that exist among the various types of statements. For example, a sentence such as "x is an artist" is called an open statement. When x is named, the sentence becomes a closed statement to which a truth value can be assigned. It should be noted that not all sentences are statements. Imperative, interrogative, and exclamatory (not shown in figure 8) sentences are not statements. Following the statement development tree from top to bottom, it is noted that not all declarative sentences are statements. Open and nonsense declarative sentences are not to be considered as statements until ambiguity is removed.

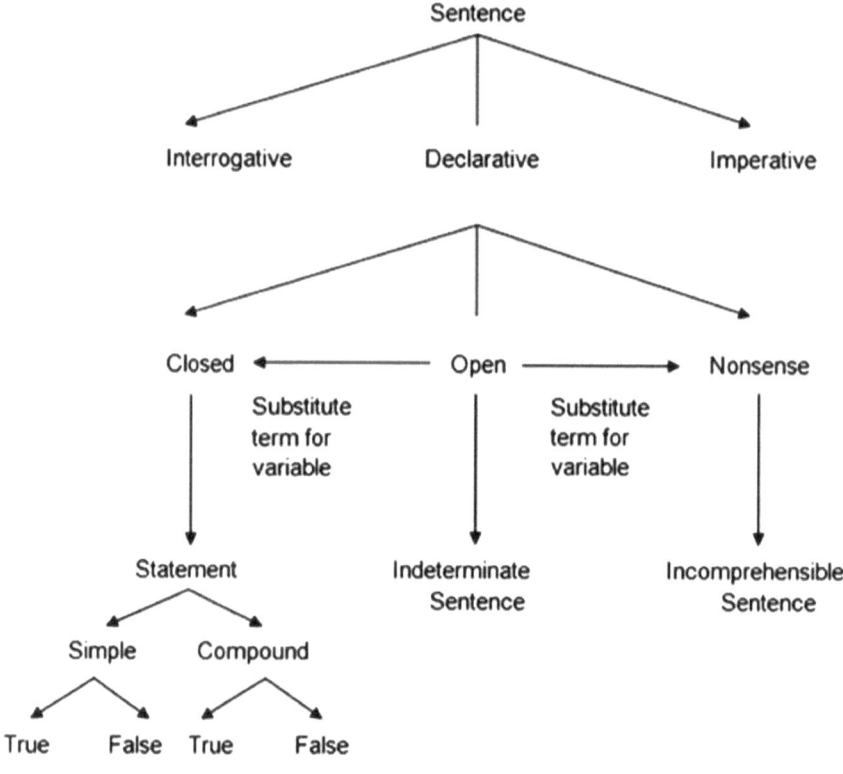

Figure 8. Statement development tree

Open sentences can be converted to nonsense sentences, indeterminate sentences, or closed statements. A closed statement results when terms yielding meaningful sentences are substituted for the variables in an open sentence. Otherwise, it results in either an indeterminate or an incomprehensible nonsense sentence. Closed meaningful statements are either simple or compound. Truth values (*T* or *F*) are assigned to simple statements based on their factual content. If the statement is compound, truth tables are employed to determine the truth or falsity of the compound statement based on the truth values of its simple statements and the logical connectives used to conjoin the simple statements.

Of primary importance in logic is the construction of compound statements from simple statements. The basic logical locutions—"and," "or," "either-or," if-then," "neither-nor"—are some of the logical connectives used to combine simple statements to form compound statements.

From the following simple statements,

A. John is smart
B. Harry is dumb,

the logical connectives "and," "or," "if-then," "not," and "neither-nor" can be used to form the following compound statements:

A. John is smart and Harry is dumb.
B. John is smart or Harry is dumb.
C. John is not smart.
D. If John is smart, then Harry is dumb.
E. Neither is John smart, nor is Harry dumb.

When the simple statements A and B are assigned a value of truth or falsehood, the question arises as to the determination of the truth or falsehood of the compound statements formed from them. Truth tables provide the answer to that question.

Truth Tables

It was noted above that the truth value of a compound statement is determined solely by its logical structure and quantification. For what truth values of its simple statements is a compound statement true? A convenient format used in answering this question is the truth table. A truth table is a tabular compilation of all combinations of truth values for the constituent simple statements that are logically connected to form the compound statement. Four fundamental truth tables for the logical connectives "and," "or," "not," and "if-then" are shown as one table below. The truth table consists of two columns of truth values for the statements and four columns of truth values for the compound statements. The fundamental truth tables are used to structure truth tables for more complex statements.

A	B	A and B	A or B	Not A	If A then B
T	T	T	T	F	T
T	F	F	T	F	F
F	T	F	T	T	T
F	F	F	F	T	T

Figure 9. Composite truth table

To simplify writing complex compound statements, we introduce the following notation:

$$A \wedge B \text{ replaces } A \text{ and } B$$
$$A \vee B \text{ replaces } A \text{ or } B$$
$$\sim A \text{ replaces not } A$$
$$A \rightarrow B \text{ replaces if } A \text{ then } B$$

The following general statements can be made about the entries in the tables:

The compound statement $A \wedge B$ is true only if the statements A and B are both true.
The compound statement $A \vee B$ is false only if the statements A and B are both false.
If A is true, its negation ($\sim A$) is false.

The compound statement $A \rightarrow B$ is called a conditional statement or an implication. It is important to note that a cause and an effect need not exist between the statements A and B. One is free to conjoin any two statements to form a conditional statement. For example, the conditional statement "If Mary goes swimming, then the tide will change" conjoins the following unrelated statements:

A. Mary goes swimming.
B. The tide will change.

The notation for the implication connective is an arrow ($\rightarrow$). The notation $A \rightarrow B$ reads, "The premise (A) implies the conclusion (B)." Another commonly used form is "If A, then B." Occasionally the "then" of the "if-then" is dropped: "If it rains tonight, the football game will be called off." At times, you will see "A only if B" used in place of "If A, then B": "It will rain tonight only if the game is called off."

The entries in truth tables for the connectives "and" and "or" were previously discussed. We now discuss the rationale behind the entries to the truth table for the implication connective ($\rightarrow$). The truth table is repeated below:

A	B	A $\rightarrow$ B
T	T	T
T	F	F
F	T	T
F	F	T

Figure 10. Implication truth table

Consider the conditional statement, "If it rains tonight, then the game will be called off." It is possible to combine the two simple statements in four ways using ^ as the connective.

It rained tonight, and the game was called off.
It rained tonight, and the game was not called off.
It did not rain tonight, and the game was called off.
It did not rain tonight, and the game was not called off.

The first combination agrees with the first row of the truth table. If it is assumed that it rained and the game was called off, realization of the occurrence, regardless of the cause, is enough to accept the entire statement as true. However, if it rained and the game was not called off, this would contradict the intended meaning of the compound statement. Hence, the compound statement would be accepted as false. This is in agreement with the second row of the truth table: if it did not rain, the nonoccurrence of rain would have no effect on whether the game was played. Hence, the compound statement would be accepted as true in both cases, corresponding to the third and fourth row of the truth table. In an implication, a true premise cannot yield a false conclusion. It is this fact that leads to this statement: in implication, a true hypothesis always leads to a true conclusion. Stated in another way, in an implication, it is false that the premise may be true and the conclusion false. In other words, $A \rightarrow B$ and $\sim(A \wedge \sim B)$ are equivalent statements; i.e., they have identical truth tables. This fact is illustrated in the following truth table:

A	B	A → B	~B	A ∧ ~B	~(A ∧ ~B)
T	T	T	F	F	T
T	F	F	T	T	F
F	T	T	F	F	T
F	F	T	T	F	T

Figure 11. Equivalent statements

In the first two columns, we list all the possible combinations of truth values for the statements A and B. In the third column, we copy the truth values of $A \rightarrow B$ from the truth table for implication. In the fourth column, we enter the truth value for $\sim B$. In the fifth column, we list the truth values for the conjunction $A \sim B$ using the truth values from columns 1 and 4. Comparing column 6 with column 3, we note that their truth values agree for each row entry of truth values for A and B. Hence, the statements heading columns 3 and 6 are equivalent.

The implication statement $A \rightarrow B$ is usually associated with three other statements commonly used in science and mathematics.

$$B \rightarrow A \quad \text{(converse)}$$
$$\sim A \rightarrow \sim B \quad \text{(inverse)}$$
$$\sim B \rightarrow \sim A \quad \text{(contrapositive)}$$

The relationships that exist among the four compound statements are identified best through the use of a truth table.

Variable Statements		Direct Implication	Converse	Inverse	Contrapositive
A	B	A → B	B → A	~A → ~B	~B → ~A
T	T	T	T	T	T
T	F	F	T	T	F
F	T	T	F	F	T
F	F	T	T	T	T

Figure 12. Associated statements

Examples are the following:

If John is rich, then Mary is poor. (Implication)
If Mary is poor, then John is rich. (Converse)
If John is not rich, then Mary is not poor. (Inverse)
If Mary is not poor, then John is not rich. (Contrapositive)

Note that the truth values listed in the third and sixth columns are identical, as are those listed in the fourth and fifth columns. Thus, the statements $A \rightarrow B$ and $\sim B \rightarrow \sim A$ are equivalent, as are $B \rightarrow A$ and $\sim A \rightarrow \sim B$. In discourse, any statement can be substituted for its equivalent. Substituting the contrapositive for the direct implication is widely used in constructing proofs for theorems in mathematics. This type of proof is called an indirect proof, or reductio ad absurdum. It was used in an appendix to show that the square root of 2 is not a rational number.

Conditional

Another connective of importance arises when the direct implication $A \rightarrow B$ and its converse, $B \rightarrow A$, are simultaneously true. The new connective, called the biconditional and denoted by $A \leftrightarrow B$, has the following truth table:

A	B	A ↔ B	(A → B) ∧ (B → A)
T	T	T	T
T	F	F	F
F	T	F	F
F	F	T	T

Figure 13. Biconditional statements

Normally, the direct implication is written as "John is rich if Mary is poor." The converse statement is written as "John is rich only if Mary is poor." The biconditional is written as "John is rich if and only if Mary is poor." Another written form for the biconditional is this: "In order that John be rich, it is necessary and sufficient that Mary be poor." In a biconditional statement $A \leftrightarrow B$, the statement $A \rightarrow B$ is translated as "A is sufficient for B, or B is necessary for A." The converse statement $B \rightarrow A$ is read as "A is necessary for B, and B is sufficient for A."

A condition is considered necessary for the occurrence of a specific event if the event will not occur in the absence of the condition. For example, a fire will not occur if oxygen is not present. The presence of oxygen is considered a necessary condition for the occurrence of a fire.

A condition is considered to be sufficient for the occurrence of a specific event if the event must occur in the presence of the condition. For example, for most substances, for a fire to occur is the attainment of a specific temperature in the presence of oxygen. Note that the presence of oxygen or the attainment of a specific temperature alone does not guarantee combustion. Each is a necessary condition. Together, they constitute a sufficient condition.

Valid Arguments

This section discusses one of the most important applications of logic, i.e., its use to determine the validity of reasoning or arguments. Up to this section, statements have been considered as abstract elements, which assume values of truth or falsehood. In this section, emphasis will be placed on examining the implication and equivalence relationships between statements. The relationship of an implication ($A \rightarrow B$) between two statements A and B is characterized by its truth table. The table shows that in an implication, B must be true whenever A is true. Stated differently, B is a consequence of, follows from, or is deductible from A.

As an example, consider the implication "If Mary studies, then she will pass." If we also assume that Mary does study—i.e., "Mary studies" is a true statement—then from the truth table, for implication, we must conclude that Mary will pass. The structure of the argument or reasoning assumes the following form:

If Mary studies, then she will pass. ($A \rightarrow B$)
Mary studies. (A is true.)
∴ Mary will pass. (B is true.)

Note that the argument contained in the example consists of two hypotheses and a conclusion. The argument assumes the form of the compound statement:

$$[(A \rightarrow B) \wedge A] \rightarrow B$$

The truth table for the above statement follows.

A	B	(A → B)	[(A → B) ∧ A]	[(A → B) ∧ A] → B
T	T	T	T	T
T	F	F	F	T
F	T	T	F	T
F	F	T	F	T

Figure 14. Valid arguments

Since the last column of the truth table contains all *T*s, the compound statement is true regardless of the truth value of the statements A and B. This states that if B is assumed to be a consequence of A and A is assumed to be true, then the statement B must be true. Any argument of this form is a valid argument. If it is stated that $A \rightarrow B$ and that B is false, what can be said about A? Can it be concluded that A must also be false? The pattern of the argument may be written as such:

$A \rightarrow B$ (A implies B.)
~ B (B is false.)
~ A (Therefore, A is false.)

Or in the form of a compound statement:

$$\{(A \rightarrow B) \wedge \sim B\} \rightarrow \sim A.$$

The truth table for the compound statement contains all *T*s. Hence, the argument is valid. Thus, in a valid argument, if the conclusion is false, the premise must also be false.

PLAUSIBLE REASONING

Introduction

IN THE PRECEDING sections, an introduction to sets and logic was presented. Sets were used to define the objects, subjects, or statements under discussion. Also stressed was the use of logic to determine the truth value of compound statements and the validity of arguments when the statements involved were true or false. Axiom systems, sets, and logic serve as part of the foundations for mathematics. Mathematics requires that the reasoning be valid and statements be true or false.

In real life, we deal with statements that cannot be determined as either true or false. Anyone interested in the legal system, politics, the stock market, or advertising will attest to the necessity of having some means to determine the credibility of statements and the plausibility of arguments. The need is paramount when uncertainty exists in either the truth value of statements or in the validity of the implications (or argument). This section on plausible reasoning describes methods for evaluating statements and arguments that will assist in the following:

1. Establishing the credibility (level of truth) of information
2. Establishing the plausibility (level of validity) of arguments $A \to B$
3. Analyzing the credibility relationships that exist between premise and consequence in valid arguments when the premise and/or consequence are credible but not specifically true or false
4. Analyzing the credibility relationships that exist between premise and consequence in other than valid arguments when the premise and/or consequence are true, false, or credible
5. Determining the credibility of statements on the basis of their substantive ties to other statements of known credibility levels

Thus, this section provides a transition from the formal structures and methods of logic, as used in mathematics, to the less formal structures and methods of plausible reasoning, as used in everyday-life situations. The section contains examples that illustrate how to apply the different forms of plausible reasoning to the analysis of everyday decision making.

Report Credibility

In practice, information is obtained from a variety of sources, including sources that are overt and covert. Rumors, expert opinion, political positions, news reports, stock market recommendations, etc., clutter the information channels such as the Internet, television news, and daily newspapers. These primary sources attempt, when possible, to give an evaluation of the reliability of the information obtained from their secondary sources. They use various meanings for the words "reliable," "possible," and "probable" to evaluate the credibility of their information and source. You just don't know what they mean when they say that an event is possible. As a reader, you can only surmise that the event may happen. The word "possible" connotes a wide range of possibilities. To sharpen the meaning of words that have a wide range of meanings, we are suggesting the use of a credibility code as shown in figure 15. This credibility code is recommended as a standard and will be used as such in the remainder of this monograph. We will use this code as a guide to evaluating the credibility of the information source and the information provided by the source. The credibility code is generally used to evaluate the reliability of the information source and the credibility (accuracy) of the information contained in the report. The information in the report can be a simple statement such as "The witness stated that he saw the defendant around 10:00 PM," or it can be the complete statement by the witness during the trial. Thus, the credibility code reflects a macroevaluation of the source of the report and its information *content*.

	Source		Information
A.	Completely reliable	1.	Confirmed by other
B.	Usually reliable	2.	Probably true
C.	Fairly reliable	3.	Possibly true
D.	Not usually reliable	4.	Doubtful
E.	Unreliable	5.	Highly unlikely
F.	Do not know	6.	Cannot be judged

Figure 15. Credibility code

This example shows how the credibility code can be used in a military situation. The military analyst receives an estimate of a situation from a field operation. The field operator uses the credibility code to evaluate his information report (B4). The evaluation B4 means that the source (person, investigator, photo, defector, etc.) is usually reliable (B), but in this instance, the information the source provided is doubtful (4). A typical information report that evaluated B4 may say that an important diplomat has been seen in a specific city. The diplomat's mission and the people he visited may be listed in the collector's report. On receiving the report, the military analyst must decide whether he will accept the evaluation provided by the originator of the report. He formulates the following questions:

1. How good is the field operator's evaluation?
2. Was the diplomat's mission accurately assessed by the source at level 4?
3. Was the source friendly or an enemy plant?
4. How reliable have been the field operator's previous reports?

The analyst cannot take the collector's evaluations at face value. He must know his sources and their biases. Knowing these, he can modify their evaluations accordingly. Techniques used to evaluate the reliability of a source are discussed in later sections.

Information Synthesis

As intelligent individuals, we are all information analysts. As intelligent individuals, we subconsciously analyze and evaluate the credibility of each piece of information we receive. To do so effectively, we have to have the means to examine the plausibility of possible relationships among the items of information we process. We should know how to synthesize seemingly unrelated items of information into an incomplete whole. We have to ask the following questions:

1. Are the information items connected?
2. Is one item a consequence of another?
3. Are there any cause-effect relationships?
4. Is the information item necessary to the conclusion of the analysis? Is it sufficient?

Further, we should know how to determine whether the relationships (vis-à-vis cause-effect, stimulus-response, input-output of information) are plausible and complete. If these relationships are lacking, we have to know how to obtain additional information in order that we may generate and test more hypotheses. It is in the objective or subjective testing of hypotheses that we as analysts exercise plausible reasoning to the greatest degree.

Information Credibility

As analysts, we deal with information and with the relationships of how the information items are connected. Varying degrees of uncertainty exist in the information items and in their interrelationships. In most practical information evaluation situations, information has a continuous range of credibility. Each statement or item of information can be evaluated as false or true or can range over a scale of credibility between true and false. One such scale of evaluation of information is suggested in a modified version of the Kent chart reproduced as figure 16. Note that this scale applies to the credibility of information items, which are simple statements. These information items may be part of a complete conversation, testimony, or status report. Thus, the Kent chart reflects a microevaluation of the information content.

Numerical Value				Nominal Values	
Average		Range		Range	
For	Against	For	Against	For	Against
1.00	0	1.00	0	Certain	Impossible
0.93	0.07	0.86–0.99	0.01–0.14	Almost certain, highly likely	Almost impossible, high unlikely
0.70	0.30	0.55–0.85	0.15–0.45	Probably likely, we believe	Doubtful, unlikely, improbable
0.50	0.50	0.46–0.54	0.46–0.54	About an even chance, about equally likely	About equally unlikely, about an even chance
0.30	0.70	0.15–0.45	0.55–0.85	Doubtful, unlikely, improbable	Probably likely, we believe
0.07	0.93	0.01–0.14	0.86–0.99	Almost impossible, highly unlikely	Almost certain, highly likely
0	1.00	0	1.00	Impossible	Certain

Figure 16. Modified Kent chart
From a privately circulated memo titled "Words of Estimative Probability" by Dr. Sherman Kent

The first line (row) of the chart indicates the type of evaluation: numerical or nominal. The entries in the second row indicate whether the column contains average numerical values or a range of values. From the third to the tenth row, the principal columns are each subdivided into two columns. The first column of the two is headed by the word "for," which signifies "the probability for." The second column is headed by the word "against," which signifies "the probability against." These entries can be interpreted as weights or as probabilities as will be noted later in this section.

Use of Kent Chart

To illustrate the use of the Kent chart, we consider a jury deliberating the testimony given by witnesses at a murder trial. The foreman asks each member of the jury to evaluate the testimony of each witness in turn. Each juror will use the terms "possible," "probable," and/or "certain" in his or her evaluation. However, they will not give an indication as to what they mean when they say that a witness's testimony was possibly false. Saying that something is possibly false can mean that it has one chance in a million of being false, or it has ninety-nine chances out of a hundred of being false. When each juror's opinion may carry the same range of uncertainty in meaning, you can see why juries deliberate over days and sometimes can't reach a verdict.

Now imagine the same jury schooled in the use of the credibility scale and the Kent chart. The odds are highly likely (my evaluation) that the jury would reach a verdict in considerably less time than one not having been properly instructed. The use of the credibility code and the Kent chart places all members of the jury on the same page as to the meaning of the words they use to evaluate a witness's testimony. The Kent chart provides the jurors with the means to assert their opinions in a more objective, nominal manner than a subjective verbal manner. It becomes easier for the jurors to understand one another and thus reach a common conclusion.

Weather forecasters are oftentimes asked to assess the possibility of rain occurring over a period of time. For example, a typical question is "What is the possibility that County X will have rain tomorrow?" Some forecasters would say that it is possible or it is clearly possible or there is a good possibility. Other forecasters will use the same terms but will state the possibility in terms of a percentage, like "It is possible with an

80% chance of rain." If the forecaster simply states that rain is possible, you have greater uncertainty in deciding whether to wear a raincoat. If the forecaster states that there is an 80% possibility for rain, you will surely wear a raincoat or bring an umbrella with you.

The examples above illustrate how the use of the Kent chart helps to achieve agreement among its users as to the meaning of terms used in the chart. Also, the Kent chart averts the confusion caused by modifying the word "possible." The nominal values assigned to the subdivisions of the realm of possibility avoid the use of the word "possible," which carries with it the complete realm of possibility as identified in the Kent chart. It is recommended that the term "possible" not be modified to indicate various levels of probability. Modification of the terms "likely," "probably," and "doubtful" is preferred and should lead to more uniform usage and understanding.

In employing the Kent chart, the chart user must be careful in deciding whether his or her experience with the source and content of his information is sufficient to assign verbal or numeric values to the information item at hand. For example, the user may have received twenty reports on the grades achieved by students taking standard tests on science. Of the twenty reports, sixteen indicate that the student received a passing grade. Based on this information, the user will assign the probability of 0.8 to the statement "Students will pass the standard test." The 0.8 figure was obtained from the ratio of relative frequency of passing grades (sixteen) to the total number of tests completed (twenty).

Continuing the example, suppose that the chart user receives only one report from the test monitor, which states that many students took the test and that most of the students passed the test. In this case, the chart user must evaluate the credibility of the test monitor based on past performances, and the user must assess what the test monitor means when he or she uses the terms "most" and "many." If the test monitor reports are highly credible, the chart user's interpretation of "most" could be "almost certain," or it could be "probably." Thus, the user will state that students are almost certain (0.86–0.99) to pass the test or that the students will probably (0.55–0.85) pass the test. What statement will be used will depend on the chart user's interpretation of the monitor's usage of "most" and "many."

Credibility Code		Kent Chart						
		Numerical Values				Verbal Values		
		Average		Range				
Source	Information	For	Against	For	Against	For		Against
Completely reliable (A)	Confirmed by others (1)	1.00	0	1.00	0	Certain		Impossible
Usually reliable (B)	Probably true (2)	0.93	0.07	0.86–0.99	0.01–0.14	Almost certain, highly likely		Almost impossible, highly unlikely
Fairly reliable (C)	Possibly true (3)	0.70	0.30	0.55–0.85	0.15–0.45	Probably likely, we believe		Doubtful, unlikely, improbable
		0.50	0.50	0.46–0.54	0.46–0.54	About an even chance, about equally likely		About equally unlikely, about an even chance
Not usually reliable (D)	Doubtful (4)	0.30	0.70	0.15–0.45	0.55–0.85	Doubtful, unlikely, improbable		Probably likely, we believe
Unreliable (E)	Highly unlikely (5)	0.07	0.93	0.01–0.14	0.86–0.99	Almost impossible, highly unlikely		Almost certain, highly likely
		0	1.00	0	1.00	Impossible		Certain

Realm of possibility (brace spanning rows 2–6)

Figure 17. Combined charts

There are many types of statements whose credibility cannot be quantified through use of the Kent chart. Specifically, general statements—such as "Within five years, a major conflict between the United States and Russia is inevitable"—are outside the realm of evaluation by the Kent chart. Many of the premises on which this general statement were based might be evaluated in terms of the Kent chart and, hence, contribute to a more meaningful evaluation of the credibility of the general statement.

At this point, it is informative to combine the charts given in figures 15 and 16 into one chart and ascertain the interrelationships that exist between the credibility of new information reports as given by the credibility code and the probability or credibility of information (facts, statements, events) as given by the modified Kent chart. The combination is given in figure 17. However, keep in mind that the section on credibility applies to sources and reports (macro) and the Kent chart applies to information items (micro).

Interchangeability

In this discussion, credibility and probability will be used somewhat synonymously. The chart given in figure 18 indicates the numeric ranges and the approximate interchangeability of the two terms. Note and compare the names attached to the various credibility levels. Note also that in the credibility code, the use of "certain" and "impossible" is avoided. This provides a hedge for human error and the unexpected.

The abbreviations SLC and SMC, used in figures 18 and 19, have the following meanings: In the middle of the credibility scale, the credibility level is designated as even. As the credibility of a statement is lowered beyond the 0.45 point at the lower end of the even level, the statement is called less credible, or LC. If the credibility is lowered beyond the 0.15 end point of the LC region, the statement is called somewhat less credible, or SLC. Comparable meanings are applied to the regions above the even region. The key point to note here is that the terms LC, SLC, MC, and SMC are credibility assessments relative to midrange of statement credibility (even).

Credibility	Credibility Code	Probability	
False	None	0	Impossible
Somewhat less credible (SLC)	Unreliable	0.01–0.14	Highly unlikely
Less credible (LC)	Not usually reliable	0.15–0.45	Improbable
Even (E)	Fairly reliable	0.46–0.54	Equally likely
More credible (MC)	Usually reliable	0.55–0.85	Probable
Somewhat more credible (SMC)	Completely reliable	0.86–0.99	Highly likely
True (T)	None	1.00	Certain

Figure 18. Interchangeability

Credibility Scale

Few information reports contain information items that can be evaluated as absolutely true or false. Many information items are incomplete, subject to error, or deceptive. Such items must be assessed as having only limited credibility (if statements) or as having a probability of occurring (if events). Because of this uncertainty in the credibility of a statement or in the occurrence of an event, it is convenient to think in terms of a scale of credibility as indicated in figure 19. In the figure, the A scale refers to a statement or event designated by A and its levels of credibility or probability. Starting from the left, we note the letter F (signifying false) designating the left end point of the scale. At this point, A has truth value (credibility) of F or zero probability. Proceeding from the left to right, the credibility of A increases from F to higher levels until the right end point of the A scale is reached. At this point, the credibility of A is T. Between two end points, A assumes credibility levels ranging from SLC to SMC. The probability ranges for A change similarly from 0.01–0.14 to 0.86–0.99.

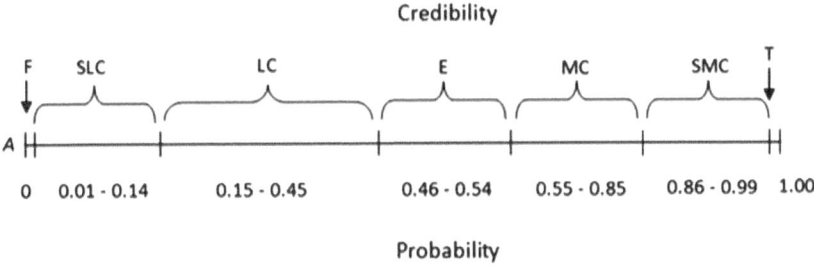

Figure 19. Credibility scale

Relational Evaluation

In this section, we consider the credibility of two statements simultaneously. To do so, we need to align two credibility scales (one for each statement) and make direct comparisons as indicated in figure 20. In figure 20, the arrows over the *A* scale and under the *B* scale indicate the level of credibility assigned to statements *A* and *B*. On the scales shown, *A* is assessed in the LC region and *B* in the MC region.

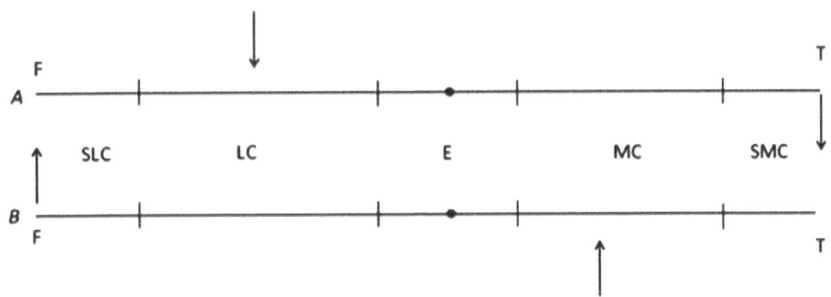

Figure 20. Paired credibility scales

Through the use of the paired credibility scales, it is possible to assess the credibility of statements, which are related through direct implication. Returning to the formal logic of statements, we recall the following two patterns of the deductive argument:

Pattern 1: $A \to B$ Pattern 2: $A \to B$

$\underline{B \text{ is false}}$ $\underline{A \text{ is true}}$
∴ A is false   ∴ B is true

Pattern 1 states that in a deductive argument, if the consequence (B) of an implication is false, then the premise (A) of the implication must be false. Referring to the paired credibility scales, pattern 1 has the truth value of B at the F position of the B credibility scale and assigns the credibility value F to statement A in the implication $A \to B$. We denote this by placing an arrow (implication) between the two scales at the F level pointing from the B scale to the A scale.

Pattern 2 states that in a deductive argument, if the premise (A) of an implication is true, then the consequence (B) of the implication must be true. Referring to the paired credibility scales, pattern 2 places the truth value of A at the T position of the A credibility scale and assigns the credibility value T to statement B in the implication $A \to B$. We denote this by placing an arrow (implication) between the two scales at the T level pointing from the A scale to the B scale. Thus, in dealing with credibility in the evaluation of the truth value of statements, the end points of the paired credibility scales yield the credibility relationships between the premise and consequence in a deductive argument whenever the extreme values of T or F can be assigned to the statements A or B respectively.

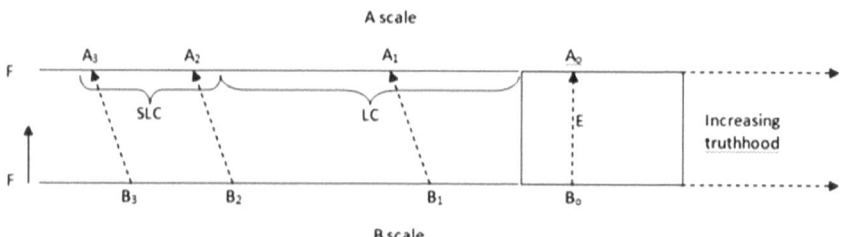

Figure 21. Credibility relationships

When the credibility of statements A or B is other than a definitive T or F in the deductive argument patterns 1 or 2, the paired credibility scales can still be employed gainfully in understanding the credibility relationships that exist between the premise and consequence.

Starting with the deductive argument pattern 1, let us assume that the credibility of statement B is E (even). Now let us observe how the credibility of statement A might vary as the credibility of statement B varies from E to F. From figure 21, it is noted that as the truth value of B tends toward F on the B scale, it cannot achieve the value of F. The bottom line (B) is shorter than the A line at the left edge. The truth value of A is not restricted to any credibility range on the A scale. It is reasonable (credible) to accept the truth value of A to be less than the truth value of B in the pattern 1 argument whenever we accept B as having a truth value other than false. This assumption is reasonably justified as follows: As the truth value of statement B tends to F from its value E, eventually, the truth value of A must tend toward F. Thus, when the truth value of B is close to F, we can reasonably expect the truth value of A to be at least as close to F as B.

Starting with truth value B_0 and progressing to the left through truth values B_1, B_2, B_3, and F for B, we can plausibly state that A will approximately have the corresponding truth values A_0, A_1, A_2, A_3, and F respectively. Note that we are not forced to place the truth values of A at the intermediate positions (A_0, A_1, A_2, A_3) shown. We are only forced to value A at position F whenever we value B at position F. We only state that as the truth value of B tends toward the value F, the truth value of A must eventually tend toward F. The closer the truth value of B is to F, the more credible is the statement that the truth value of A is at least as close as the truth value of B to F is.

Credibility Patterns

Consider the two statements:

A: Corporation X is attempting to purchase Corporation Y.
B: Corporation X is secretly purchasing stocks of Corporation Y.

First, assume that every time Corporation X attempts to purchase another corporation, it subsequently purchases stock in that corporation.

On the basis of this assumption, we can state that $A \rightarrow B$ is a valid implication. Now suppose that it is difficult to obtain information on the intentions of Corporation X relative to purchasing Corporation Y. Corporation Y can, through trade sources, obtain some information on whether Corporation X is purchasing or intends to purchase its stock; that is, it is difficult for Corporation X to cover their purchases of Corporation Y stock. Hence, Corporation Y directs its resources to the collection of intelligence on Corporation X's purchases of its stock in order to determine whether Corporation X will attempt the purchase.

If intelligence indicates that stock purchases are highly likely, then Corporation Y can anticipate that a purchase attempt will be made by Corporation X. Yet there will be some reluctance on Corporation Y's part to say that Corporation X's purchase attempt is highly likely. Rather, the tendency is to have the likelihood of statement A occurring lag the likelihood of statement B occurring. If evidence continues to accumulate and reduces the likelihood of B occurring, then there is a strong tendency to evaluate the possibility of Corporation X attempting a purchase of Corporation Y at a still lower level.

At the other end of the truth scale, we have the deductive argument pattern 2 repeated as below:

$$A \rightarrow B$$

$$\underline{A \text{ is true.}}$$
$$\therefore B \text{ is true.}$$

In this deductive pattern, the truth of the premise (A) guarantees the truth of the consequence (B). Now let us investigate what can be said of the truth value of B when the truth value of A is less than T. Starting with the truth value E for statement A and allowing the truth value of A to progress from E to T on the A scale, we are forced to accept that the truth value of B on the B scale will be T when the truth value of A arrives at T on the A scale. At intermediate truth value positions (E, A_1, A_2, and A_3 for A on the A scale), it is plausible to associate values E, B_1, B_2, and B_3 for B on the B scale, as indicated in figure 11. Note that we have placed the truth values for B slightly higher than are the corresponding values for A. We are not forced to do this. We only state that it is plausible that we do so to guarantee that B will be true when

A becomes true. It is quite possible for B to be true even if A is false. Hence, it is reasonable, in cases of uncertainty, to assign B a slightly higher truth value than A.

Credibility Relationships

Consider the two statements A and B listed in the last section. Again, suppose that each time Corporation X attempts to purchase Corporation Y, it subsequently initiates a purchase of Corporation Y's stock. Hence, we can state that $A \to B$ is a valid implication. Now suppose that, through an outstanding information collection effort, Corporation Y is able to obtain information on the intentions of Corporation X relative to the purchase of Corporation Y. As evidence accumulates, supporting Corporation X's intentions, Corporation Y feels at least equally assured that Corporation X will start purchasing Corporation Y's stock. If the evidence accumulates and indicates a trend toward Corporation X reducing its efforts to purchase Corporation Y, then Corporation Y can be at least equally assured that Corporation X will stop purchasing its stock. At this point, with the assistance of figures 21 and 22, it is possible to summarize some intuitively acceptable practices in interpreting deductive patterns when the statements therein have credibility other than T or F.

Interpretation 1: When the credibility of A or B is in the E range—i.e., the truth or falsity of A or B is about even—the following patterns evolve:

$A \to B$ $\quad\quad\quad$ $A \to B$

B is E. $\quad\quad\quad$ A is E.
A is E. $\quad\quad\quad$ B is E.

The two patterns indicate that when two statements are linked together through a direct implication, and either statement has a truth value of approximately E, it is plausible to accept the truth value of the remaining statement at approximately the E level. If one interprets the E

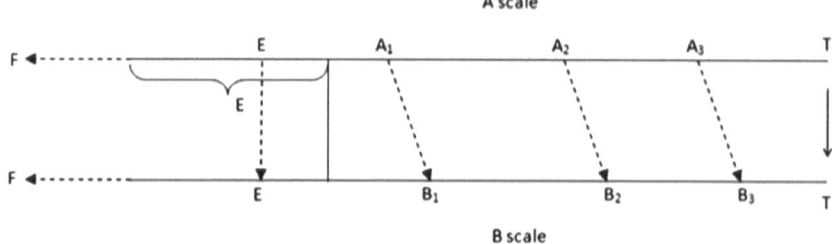

Figure 22. Credibility relationships

region as the region of uncertainty, then the two patterns state that uncertainty in premise or conclusion begets uncertainty in conclusion or premise respectively. See figure 21.

Interpretation 2: When the truth value of B tends toward F, it is plausible to conclude that the truth value of A tends slightly more rapidly than the truth value of B toward F; i.e., the credibility of A remains less than that of B as B tends toward F and assumes truth values close to F. See figure 21.

$$A \rightarrow B$$

<u>B is LC.</u>
A is SLC.

Interpretation 3: When the truth value of A tends toward T, it is plausible to conclude that the truth value of B tends slightly more rapidly than the truth value of A toward T; i.e., B remains more credible than A as A tends toward T and assumes truth values close to T. See figure 22.

$$A \rightarrow B$$

<u>A is MC.</u>
B is SMC.

Interpretation 4: It is acceptable to have A false in the implication $A \rightarrow B$ and still have B true. As A becomes more true—i.e., more evidence is gathered to support A—then based on the credibility of A, the credibility of B exceeds the credibility of A and tends more rapidly than A does toward higher levels of credibility. See figure 22.

Interpretation 5: It is acceptable to have B true in the implication $A \rightarrow B$ and still have A false. As B loses its credibility—i.e. tends toward falsehood—then based on the loss of credibility in B, the credibility of A lags B and tends more rapidly toward lower levels of credibility.

The credibility relationships obtained between A and B (as the truth values of A and B range between T and F in the implication $A \rightarrow B$) can be summarized through the use of the credibility scale diagram shown in figure 23 and the ten deductive logic patterns shown in figure 24.

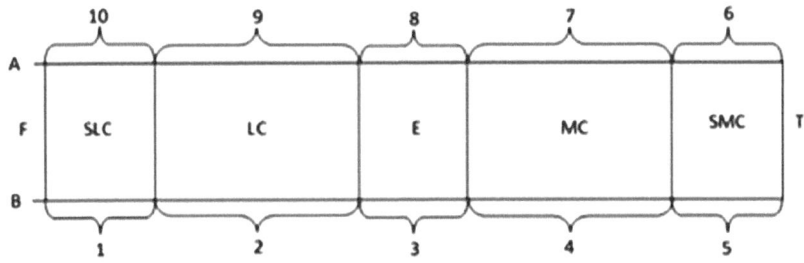

Figure 23. Credibility regions

Deductive Patterns

Again, we assume that the credibility (level of truth) value of A or B is continuously variable and can be measured between F and T through the intermediate ranges SLC, LC, E, MC, and SMC, as indicated in figure 23. The argument pattern numbers of figure 24 and the numbers in figure 23 are related as follows: The second statement in the premise involves statement B in patterns 1–5 and statement A in patterns 6–10. In each pattern, the pattern number corresponds to the credibility region in which the credibility value of A or B lies. For example, in pattern 3, B assumes a credibility value of E, which lies in region 3. In pattern 7, the credibility value of A lies in region 7.

Starting with the deductive pattern 1 in figure 24, we note that this pattern restates the previous finding that A must be false if B is false. Parenthetically, pattern 1 indicates that if the credibility value of B ranges in the SLC regions, we expect the credibility value of A to range in the same region and be slightly lower than B.

When the credibility value of *B* ranges in the LC region—as indicated by pattern 2 in figure 24 and as located in the credibility scale of figure 23—it is more difficult to state definitively what the corresponding credibility value of *A* should be. As long as we have assessed the credibility value of *B* to be less than even, the tendency is to assign a credibility value to *A* at least as low as *B*. Thus, pattern 2 indicates that when *B* is LC, *A* can plausibly be assigned credibility values from LC down through SLC.

1. $A \rightarrow B$
   <u>*B* is F (SLC).</u>
   A is F (SLC to F).

2. $A \rightarrow B$
   <u>*B* is LC.</u>
   A is LC to SLC.

3. $A \rightarrow B$
   <u>*B* is E.</u>
   A is E to LC.

4. $A \rightarrow B$
   <u>*B* is MC.</u>
   A is MC to E.

5. $A \rightarrow B$
   <u>*B* is (SMC).</u>
   A is SMC to MC.

6. $A \rightarrow B$
   <u>*A* is T (SMC).</u>
   B is T (SMC to T).

7. $A \rightarrow B$
   <u>*A* is MC.</u>
   B is MC to SMC.

8. $A \rightarrow B$
   <u>*A* is E.</u>
   B is E to MC.

9. $A \rightarrow B$
   <u>*A* is LC.</u>
   B is LC to E.

10. $A \rightarrow B$
    <u>*A* is (SLC).</u>
    B is LC to SLC.

Figure 24. Deductive patterns

Patterns 3, 4, and 5 all exhibit a lower assessment of the credibility value of the premise (*A*) relative to the assessment of the conclusion

(B), regardless of the assessed credibility value of the conclusion. This relationship persists in the implication $A \rightarrow B$ because B is the necessary condition for A. It supports the credibility of A, but it is not sufficient for (does not guarantee) the truth or existence of A. One can note in pattern 5 that it is not permissible to assign A the truth value T while B is assigned a credibility value other than T. Hence, the tendency is to have the credibility value of A lag the credibility value of B as B tends toward a truth value T.

One might call the above observation the principle of overcompensation; i.e., not having faith in the credibility of the conclusion begets less faith in the credibility of the premise from which the conclusion was derived.

Patterns 6–10 indicate a more optimistic evaluation of the credibility of the conclusions when viewed through the credibility value assigned to the premises. For example, pattern 6, which has previously been discussed, indicates that when A is assigned the truth value T, B must be assigned the truth value T. If A is SMC, then B will range in SMC to T.

The optimism persists in patterns 6–10 because the premise (A) is the sufficient condition for B; A guarantees the credibility, truth, or existence of B. One can note in pattern 10 that it is not permissible to assign a truth value of F to B while A is assigned a credibility value other than F. Hence, the tendency is to have the credibility value for B higher than the credibility value of A as A trends toward the truth value F.

In summary, the ten patterns indicate the following two general interpretations:

1. If one does not accept your conclusions, then one will be less inclined to accept the premises from which the conclusions were validly drawn.
2. If one accepts your premises, then one will be more inclined to accept the conclusions that were validly drawn from your premises.

Plausibility Synthesis

There are many situations in everyday life when one has to determine whether some news item, stock recommendation, or political statement is true, credible, or false. For example, your stock broker may make the following stock recommendation (P): purchase Company A's stock. It is producing a new Gizmo X that should drastically increase earnings. You could trust your advisor, accept the recommendation, and purchase the stock. If you did so, you would not be alone. Many stocks are purchased based solely on the advice of a stock broker or financial advisor. However, to be an informed investor, you need to seek information from other sources, which could help you make a decision to buy or pass.

What has been presented in the above sections on plausible reasoning provides you with the techniques to use to evaluate the credibility of simple implications like $A \rightarrow B$. However, statements such as statement (P) contain several premises from which to draw a conclusion. Each premise requires evaluation as to its credibility prior to its use in determining the credibility of the conclusion. This section provides you with an example on how information from several sources can be synthesized, how its plausibility can be determined, and how it can then be used to determine the plausibility of the conclusion.

We shall use the stock broker recommendation (P) as an example. Before you act on the recommendation (P), you should seek information on Company A from sources other than a single source—your stock broker. Another information source may advise that for statement P to be credible, Company A should be increasing its purchases of products U and V, noting that these products are necessary for Company A to produce Gizmo X. An advisory service may note that for statement P to be credible, you should see an increase in volume in Company A's stock. With these pieces of additional information, you should be in a better position to apply plausible reasoning and make an informed decision on the purchase of Company A's stock.

Using plausible reasoning, you would note that the information obtained from your sources can be reduced to three statements. Each of the statements can be assigned a credibility value F, SLC, LC, E, MC, SMC, or F.

Q: Producing a new gizmo will increase earnings.

R: To produce new gizmo requires purchase of products U and V.

S: Increased earning will increase stock volume.

The three statements can be combined in eight different ways: one at a time, two at a time, all or none. We shall apply the techniques of plausible reasoning to the synthesis of three statements together. The techniques also apply to situations where two of the three statements are synthesized.

It is possible to logically conjoin the three statements in four different ways:

$Q \wedge R \wedge S \rightarrow$ Purchase stock in company A.
$Q \wedge R \vee S \rightarrow$ Purchase stock in company A.
$Q \vee R \wedge S \rightarrow$ Purchase stock in company A.
$Q \vee R \vee S \rightarrow$ Purchase stock in company A.

Note that $\wedge$ = "and" and $\vee$ = "or." Using the $\wedge$ connector between statements means that the credibility of each statement shall be used to determine the credibility of the composite statement. Using the $\vee$ connector between statements means that the credibility of either or both statements shall be used to determine the credibility of the composite statement. For example, in $Q \wedge R \vee S$, the credibility of Q and R must be used, but the credibility of S may or may not be used.

In a very rare case, your credibility assessment may lead you to decide that statements Q, R, and S are true. In this case, the conjunction of the three statements is also true, and you will able to synthesize the following implication:

$Q \wedge R \wedge S \rightarrow$ Purchase stock in company A.
$Q \wedge R \wedge S$ is true.
Therefore, purchase stock in company A.

But in most cases, you will find that Q, R, and S can't all be deemed to have truth value (T). At best, the statements will have credibility values from F to SLC to LC to E to MC to SMC to T. To determine the plausibility of statement P, you are now faced with the four possible implications wherein each statement can take on any of the seven plausibility ranges. Your chore now is to determine the credibility of each of the statements (Q, R, and S) and then determine the plausibility of each of the four ways the statements Q, R, and S are conjoined.

For statement Q, you have to estimate the likelihood that the production of a new gizmo will produce an increase in Company A's earnings. For statement R, you have to verify whether Company A is purchasing products U and V. For statement S, you check the volume of Company A's stock and note any recent increases in the volume. Once you have determined a credibility range for each of the statements, your next action is to determine the plausibility of the four synthesized statements.

To lessen the effort, some rules of thumb can be used to determine the credibility of the composite statements:

Rule 1: All statements (premises in the implication) are to be sorted into two categories designated by $\wedge$ and $\vee$. The $\wedge$ category will contain those statements that appear to have the highest cause-effect relationship with the conclusion and quality of source. Statements in this category will be conjoined by $\wedge$. The $\vee$ category will contain the remaining statements. They will be conjoined by $\vee$. For example, if you sort the statements U, V, W, X, Y, and Z and place the statements X, W, and Y in the $\wedge$ category and U, V, Z in the $\vee$ category, the composite statement will take the form $X \wedge W \wedge Y \vee U \vee V \vee Z$.

Rule 2: For statements in the $\wedge$ category (i.e., conjoined by $\wedge$), the credibility range for the composite statement is obtained by averaging the numerical midpoints of the range for each statement. For example, using the credibility scale in figure 19, if in $Q \wedge R \wedge S$, Q is MC with midpoint 0.7, R is MC with midpoint 0.7, and S is SMC with midpoint 0.925, then the composite statement has midpoint $(0.7 + 0.7 + 0.925)/3 = 0.775 =$ MC.

Rule 3: For statements in the $\vee$ category (i.e., conjoined by $\vee$), each statement can be used to adjust the credibility of statements conjoined by $\wedge$, as illustrated in figures 11 and 12 titled Credibility relationships.

NUMBERS

Executive Summary

NUMBERS CAN BE considers as names for the points on the line called the *x*-axis of the Cartesian plane. Numbers are made up of two parts, a whole part and a decimal part. In typical numbers 26.345 and 27.345, the whole parts (26 and 27) identify points on that line spaced a unit apart. The decimal part (345) identifies a point between the successive whole numbers, 26 and 27. Whereas the points on the line spaced a unit apart are given different names (3, 45, –22, etc.), the same names are given to points that fill the line segment between any two successive numbers. The set of whole numbers and decimal numbers form the real number continuum.

In the monograph, we shall see how an axiom system, sets, set operations, and logic are used to define sets of numbers as subsets of the universal set of real numbers. Starting with the natural numbers (N or whole numbers), we note that the set of natural numbers has an axiomatic basis as given by Nicolas Peano. The axioms show us how the successor operation is used to count from one natural number to the next. As noted above, the natural numbers are viewed as points on the positive *x*-axis of the Cartesian plane. By adding the negative natural numbers and the number 0 to the natural numbers, we arrive at the set of integers (Z). With the integers, we are able to count indefinitely, in either direction, from integer to integer. But what about identifying and naming the decimals between the integers?

By defining the set of rational numbers (F for fractions) as the ratio of integers, p/q, $q > 0$, we are able to identify the rational decimals between the integers. The ratio p/q always results in a decimal consisting of a finite number of digits or repeating cycles of digits. When we name points in the interval between integers with rational numbers, we note that not all the points in the interval can be named. What are the unnamed points in the interval? Here is where use is made of Cantor's diagonal process to show that numbers other than the rational numbers cover points on the line segment between integers. This leads to the necessity to name the points on the line segment not covered by rational numbers. One such set of numbers is the algebraic numbers designated as A. The set (A) is identified as solutions (roots) to the general polynomial equation. Still, points on the interval remain unnamed.

In naming the remaining points, two sets of numbers that differ from the four types above are identified. These numbers have noncyclical, nonending sequences of digits (not ending in zeros). One set is the complement to the set of rational numbers. The set is called irrational numbers (I). The irrational numbers cannot be expressed as a ratio of integers. The second set is the complement to the set of algebraic numbers. This set includes numbers such as the Naperian base (e). The set is called transcendental numbers (T). Together, the six sets of numbers name all the points on the line and form the real number continuum.

At this point, the monograph digresses to introduce some special properties of the real number system. Prime numbers are identified, and Eratosthenes's sieve is used to separate them from other numbers. In keeping with the proliferation of digital technology, we describe the binary number system and show how to convert between base 2 and base 10 number systems

Questions still remain as to the relative size of the six sets of numbers. To answer the questions, the one-to-one correspondence is introduced and used to divide the six sets of numbers into two equivalence classes. The equivalence classes are identified by the cardinal numbers N_0 and c. The natural numbers, the integers, the rational numbers, and the algebraic numbers form one equivalence class with cardinal number N_0. The irrational numbers, the transcendental numbers, and the real numbers form a second equivalence class with cardinal number c. We then find that the power set, the set of all subsets of the real decimal numbers, does not have N_0 or c as its cardinal number. We denote its cardinal number f. The three cardinal numbers are related by $N_0 < c < f$, where < indicated an order according to size or inclusion.

Noting that the relationship for n finite $n < 2^n$ also held for cardinal numbers (p), we used the fact $p < 2^p$ to allow us, through exponentiation, to identify an infinite sequence of transfinite cardinal number $a_1 < a_2 < a_3 < ...$ However, this sequence did not constitute all the transfinite cardinal numbers. We were able to construct sets with cardinal numbers that were not in the sequence.

CLASSES OF NUMBERS

Zero

THE NUMBER 0 remained a mystery, an unneeded concept throughout many early civilizations. There is some evidence of its use in Babylonian and Egyptian civilizations. Ancient Greeks were more philosophical, asking, "How can we make something out of nothing?" There is evidence that by the fourth century BC, the south-central Olmec civilization used a calendar and numeral systems that contained 0 as a numeral. However, it wasn't until AD 130 that Ptolemy used 0 by itself to indicate nothing rather than being a placeholder. By AD 525, tables used with Roman numerals contained the use of a true zero, not as a numeral but as a word, "nulla," with the meaning of "nothing."

By the seventh century, the concept of 0 as a numeral gained acceptance in advancing civilizations throughout the world. Still, man only thought about positives. First, there were tallies to keep track of real objects. There was no need to tally or count zero objects. You did not have zero bananas; you had no bananas. Eventually, as wealth was created, money (instead of objects) was exchanged and served as a measure for value. The money changers and, eventually, accountants had to balance accounts. When amounts balanced out, zero became an accepted result.

The introduction of 0 as an integer brought with it some benefits and some problems. Zero could be thought of as the fulcrum that balanced the real numbers. Positive numbers to the right and negative numbers to the left on the real number balance bar or number continuum. Zero could also be thought of as a mirror at the zero point on the number axis. Positive natural numbers are the object; the negative numbers, the reflection. Adding or subtracting two numbers results in a number. That is the property of closure. When you add 5 to 7, you get 12, which is larger than 5 or 7. With physical elements, as with positive numbers, the sum is greater than any of its parts. This principle does not hold for zero. 0 + 0 = 0. 0 + 2 = 2. 2 - 0 = 2. You might say that 0 lacks mass or energy.

As a multiplier, 0 assumes the role of a vacuum. It sucks the number into itself: 5 × 0 = 0. As a divisor, 0 assumes the role of a creator, 5 ÷ 0 = what some call infinity. In mathematics, it is called a singularity. Cosmologists say that the creation of the universe resulted from a singularity. Could it be that in a prior universe, someone divided by 0, resulting in the big bang and the birth of the universe? An idea quite far-fetched.

However, in our modern-day society, division by 0 is to be avoided like the plague. This is especially true where computer programs control the operating performance of automobiles, manufacturing processes, medical procedures, financial markets, and who knows what else. If through some fault, a division by 0 occurs in the controlling program, the control process will abort, and the system will fail. Not many years ago, the USS *Yorktown*, a missile cruiser, met with that fate. A newly installed program that controlled its 80,000-horsepower propulsion system tried to divide by 0. The result was a disabled warship and the need for a return to the shipyard for repairs.

Natural Numbers (*N*)

The numbers (1, 2, 3, 4...) we use in counting are called natural numbers. We normally refer to each number as a digit or a whole number. In our elementary schools, we never really defined the term "number." By the time a child starts going to school, the child has intuitively learned to use numbers to count. They also are exposed to the term "infinity" and are taught not to use it as just another number, but that lesson does not appear to be well-learned. For example, there was an advertisement on TV where a moderator asked a group of children, "What is the largest number you know?" The children in turn reply, "a billion," then "infinity," then "infinity plus 1," then "infinity times infinity," and finally, "infinity to the infinite power." They appear to be really smart kids.

However, they and the moderator were wrong in applying arithmetic processes to numbers involving infinity. Normal arithmetic processes do not apply to quantities that are infinite. Before we deal with determining the size of sets that are infinite, we should have some idea of what we mean when we say that a set is finite or infinite. To do so, we introduce an axiom system that will define what we mean by the set of natural numbers. We then use the one-to-one correspondence to define the terms "finite" and "infinite" as they apply to the natural numbers and then to the size of any set of elements.

We are familiar with the fact that Euclid's axiom system serves as a foundation for the development of plane geometry. What is not familiar is that Nicolas Peano (1858–1932) originated an axiom system that serves as a foundation for the class N of natural numbers in the year 1889. The axiom system states that the term "number" and binary operation (s) between the members of N are undefined. The operation (s) is called successor. If x and y are members of the set of natural numbers (N), then $y\ s\ x$ states that y is the successor of x; i.e., y succeeds (follows) x. For example, 4 s 3, 16 s 15, and 5 $ssss$ 1.

The Peano Axioms

1. There's an element 1 in N such that for all y in N, 1 is not a successor of y.
2. For every element x in N, there is a unique element y in N such that $y\ s\ x$.
3. If $y\ s\ x$ and $a\ s\ b$ and $y = a$, then $x = b$.
4. If a subset (G) of N contains the element 1 and if every element of G has a successor in G, then $G = N$.

Briefly, the axioms state that N has a first element (we normally call it 1), every element of N has a unique successor, and no two elements of N have the same successor. The fourth axiom states that an important method of proof in mathematics (mathematical induction) works.

Using the axioms, we can construct the set (N). We start with the first element 1 in N. We can then order the elements of the set (N) by applying the successor operation in order to the previously obtained successor. For example s1 = 2, ss1 = 3, sss1 = 4, s4 = 5. As we proceed, we can label the elements of N in any way we please. It just happened that the labeling of the elements of N was done prior to the discovery of Peano's axioms. Therefore, staying with history, we label the first element 1, its successor as 2, and the successor of 2 as 3 and then continue applying the successor operation as far as we please.

In that the successor operation can be applied indefinitely to obtain additional elements in N, we *accept* that the set of natural numbers (N) is infinite. We assign the label N_0, or aleph null, to this level of infinity. We also *accept* that N_0 is the least number representing increasing levels of transfinite numbers. We underline the word "accept" because we still have to define what we mean by finite and infinite.

We now return to the question, "Why is the set of natural numbers (N) deemed to be infinite?" To answer the question, we have to define what we mean by infinite. We have the option to define infinite first and then define what we mean by finite. Or we can do so in reverse order. Since we normally deal with finite numbers, it seems more natural to define finite and then define infinite in terms of finite. Here is where the one-to-one correspondence between elements of two or more sets becomes an important tool in understanding how infinities are classified.

We state that a set (S) is finite if it is empty or its elements can be placed in a one-to-one correspondence with a (proper) subset of the natural numbers (N). For example, the set (S) containing the twenty-six letters (*a, b, c, d*...) of the alphabet is finite because it is possible to pair natural numbers (1, 2, 3, 4...26) with the elements of S. The pairing goes (1, *a*); (2, *b*); (3, *c*) . . . until we reach the natural number $n = 26$, yielding (26, *z*). We can also view the pairing as follows:

$$1 \quad 2 \quad 3 \quad 4 \quad 5 \quad 6 \quad 7 \quad 8 \quad 9 \quad 10 \ldots$$
$$A \quad B \quad C \quad D \quad E \quad F \quad G \quad H \quad I \quad J \ldots$$

However, there are instances where it is not possible to specify the natural number (*n*) in the definition. Take for example the number of stars in the Milky Way. Another example is the number of grains of sand on earth. Due to the effect of wind, tide, and erosion on sand, the number of grains of sand (*n*) is large but variable. However, we are certain that the grains can be counted at any moment in time. We just accept the number (*n*) as finite although variable.

We now state that a set (S) is infinite if it is not finite. The definition is simple in statement, but its implications are complex. One interpretation of a set (S) being infinite is the nonexistence of a one-to-one correspondence between the elements of S and a subset of natural numbers (N) containing n numbers. Another and more useful definition for a set (S) to be infinite is that it will be possible to obtain a one-to-one correspondence between S and a proper subset S' of S. For example, the set of natural numbers (N) is infinite because it has proper subsets (for example, the even numbers $2n$ and odd numbers $2n + 1$) that can be placed in a one-to-one correspondence with the elements of N. The correspondences are $(n \leftrightarrow 2n)$ and $(n \leftrightarrow 2n - 1)$.

Integers (Z)

As shown in figure 2, the natural numbers, also known as counting numbers, are normally visualized as names for points that are placed a unit apart along the positive x-axis on the Cartesian plane. The negatives of the natural numbers and 0 are added to the natural numbers to form the set of numbers we call the integers (Z). You can think of the number 0 as the fulcrum or balancing point for the integers. Together, a unit apart, the integers span the entire x-axis.

Since the integers consist of negative numbers, the notion of the integer is a late comer in the history of numbers. Like 0, negative numbers remained an unacceptable concept until the seventeenth century. Prior to then, there was some use of negative numbers to represent debts. Otherwise, negative numbers resulting from any calculations were deemed to be unacceptable and discarded as absurd.

Decimal Numbers (D)

We know that between successive integers, say n and $n + 1$, there exists a continuum of numbers that are added to n. For example, starting at the natural number 2, numbers ranging from 0 to 0.9999..., are added to 2 until they approach the number $3 = 2.9999...$. Thus, to every point on the x-axis, there is a composite number consisting of an integral part and a decimal part. These numbers can be represented by a number in the form

$$...n_4 n_3 n_2 n_1 . a_1 a_2 a_3 a_4 ...,$$

where each n_i and a_i is a digit or 0. The n_i represents the integral part of the composite number. The a_i represents the decimal part of the composite number. For example, in the number 478.234, the 478 is the integral part, and the 0.234 is the decimal part. Accepting that the decimal part, 0.0 to 0.9999..., of a composite number, is the same between all successive integers, we no longer need be concerned about the structure of the integral part of a number. Our main interest lies in the structure of the decimal part. Through the decimal part, we can identify whether the number is real or rational.

Rational Numbers (*F*)

The rational numbers (*F* for fractions) are numbers that can be expressed as ratios or fractions as in *p*/*q*, where *p and q* are integers and *q* is not 0. When we divide the denominator into the numerator, the resulting number contains an integral part and a decimal part. The decimal part of a rational number either terminates or results in a cyclical group of digits. The number of digits in the cycle is called the period of the decimal. Examples are the following:

$\frac{1}{8}$ = 0.125	Period 0
$\frac{3}{11}$ = 0.272727...	Period 2
$\frac{17}{9}$ = 1.88888...	Period 1
$\frac{3}{7}$ = 0.428571428571...	Period 6

At each stage of the division process for *p*/*q*, the remainder assumes a value less than *q*, which is the value of the denominator. For example, try dividing 3 by 7 (3/7). The remainder values are 2, 6, 4, 5, 1, 3—all less than 7, the denominator. This sequence of remainder values repeat cyclically when the division process continues, resulting in cycles of that length in the answer to the division, for example,

$$\frac{3}{7} = 0.428571428571..., \text{ period 6.}$$

The rational numbers (*F*) have a unique property. Let an unending sequence of zeros be appended to the decimal part of each terminating rational number. The addition of zeros makes the natural number a nonterminating type. We can make the following statement: there is a one-to-one correspondence between unique periodic decimals and rational numbers. That is a very strong statement. For those interested, the processes used to convert cyclical decimals to ratios (rational numbers), and vice versa, are described in appendix 1.

Let us return to our representation of rational numbers (F) as points on the x-axis of the Cartesian plane. Let us also restrict our attention to a line segment on the axis of length 1 (between two adjacent natural numbers). We may ask the question, "How many points on the line segment are represented by rational numbers?" The simple answer is an infinite number of points. However, how do we know that?

First, we can show (see appendix 2) that the rational numbers can be placed in one-to-one correspondence with the natural numbers; i.e., they have the same cardinal number N_0. That knowledge does not give us any information on the distribution of rational numbers between two adjacent natural numbers. To illustrate how close together the rational are distributed on a line segment of unit length, we compute the average for two rational numbers, r_1 and r_2, representing two different points on the line segment.

The average of the two rational numbers is given by $(r_1 + r_2)/2$. The result is a rational number (r_3) located at a point halfway between the two original points. When we repeat the process, averaging r_3 with both r_2 and r_1, the result is two more rational numbers. We can continue this averaging process until the nth stage of averaging is reached. At the nth stage, 2^{n-1} new rational numbers are formed. All these rational numbers lie between the original two rational numbers. The averaging process can be continued indefinitely. This leads to the conclusion that however small the distance between two rational numbers, it is always possible to construct a new rational number between the two. Stated another way, there is an infinity of rational numbers between any two distinct rational numbers. Yet if we were to measure the length of the unit interval occupied by the rational numbers, we would find that measure to be 0.

Algebraic Numbers (A)

From the preceding paragraph, it appears that there is no room in a unit interval for any numbers other than rational numbers. That is very far from the truth. There exists a set (A) of numbers, known as algebraic numbers, consisting of the solutions (roots) to the general polynomial equation as given below:

$$a_0 x^n + a_1 x^{n-1} + a_2 x^{n-2} + \cdots + a_{n-1} x + a_n = 0$$

In this equation, n is a natural number, the a_i are integers, and a_0 is greater than 0. An example of this polynomial is the well-known quadratic equation $ax^2 + bx + c = 0$. Its solutions (roots), in terms of a, b and c, are given as

$$x = -b \pm [\sqrt{(b^2 - 4ac)}]/2a$$

A simpler example is the equation $(q) x = p$. The root $x = p/q$ of this algebraic equation is a rational number for all integral values of p and q, q not 0. Therefore, the rational numbers are a subset of algebraic numbers. However, the polynomial $x^2 - 2 = 0$ has solutions (roots) $x = \pm\sqrt{2}$ that are not rational numbers. This tells us that the rational numbers are a proper subset of the set (A) of algebraic numbers. Yet as indicated in appendix 3, there exists a one-to-one correspondence between the algebraic numbers and the natural numbers.

Real Numbers (R)

The real numbers, denoted R, consists of all numbers that can be written as an unending decimal of the form,

$$b_1 b_2 b_3 \ldots b_m . a_1 a_2 a_3 \ldots a_n \ldots,$$

where the a_i and b_i are digits 0 through 9. The decimal part is cyclical or continues indefinitely; i.e., it does not end in zeros. The set of real number (R) form what is called the real number continuum. Graphically, all the points on the x-axis represent the real number continuum. However, it is the structure of the decimal part of the number that determines whether the real number is a rational or other type of number. Examples of real numbers are as follows.

The natural numbers are real numbers since each has a decimal representation in the form.

$$3 = 2.99999...$$
$$26 = 25.99999...$$

The rational numbers are real numbers since they can be written as unending sequences.

$$\tfrac{1}{9} = 0.11111...$$
$$\tfrac{11}{3} = 3.66666...$$

The algebraic numbers are real numbers since they are roots of the general polynomial equation. (Example: $ax^2 + bx + c = 0$).

$$\text{The irrationals, pi, } \sqrt{2}$$

The question now is "Do the real numbers have a cardinal number other than N_0?" The real numbers (R) contain as subsets all the number sets described previously. In set notation,

$$N \subset F \subset A \subset R.$$

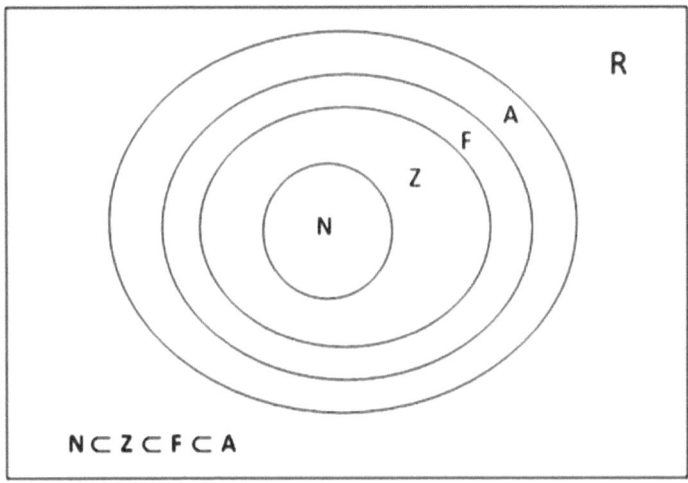

Figure 25. Subsets of R with cardinal number N_0

Do the real numbers still have cardinal number N_0? One would think that the real numbers should rank higher in cardinal number than N_0. Georg Cantor, using his diagonal procedure, proved that real numbers do not have the cardinal number N_0. We could leave it at that. However, here we have a great opportunity to see how sets and logic enter into the proof that the set (R) does not have cardinal number N_0. The structure of the logical argument is as follows:

$$A \rightarrow B \qquad\qquad A \rightarrow B$$

$$\underline{A \text{ is true}}; \qquad\qquad \underline{B \text{ is false}}.$$
$$\therefore B \text{ is false}. \qquad\qquad \therefore A \text{ is false}.$$

The statements A and B are as follows:

A: R is the set of real numbers.
B: R does not have cardinal number N_0.

The argument on the left is the direct implication. The argument on the right is the contrapositive. The contrapositive argument is called the indirect proof or reductio ad absurdum.

Georg Cantor (1845–1918)) used the indirect method of proof. He assumed that the statement B is false; i.e., the real numbers (R) between 0 and 1 had the same cardinal number N_0 as the natural numbers. As such, he was able to list the decimals and use the one-to-one correspondence to pair each decimal number with a natural number. What Cantor did was equivalent to placing the decimal digits of the real numbers (r_i) into the squares of the array of figure 26. The cubes in the array are filled with the digits of the first six real numbers.

Cantor entered the digits of the real number into the array, knowing that the listing of numbers into the array could never be completed. He labeled each digit in a cube of the array according to its position in the array. For example, the digit in the cube located at the intersection of the second row and third column would be labeled a_{23}. In the figure, it would be $a_{23} = 3$. Digits in cubes located at the intersection of rows and columns with the same subscript lie on the diagonal of the array. These digits, in order 372211, are bold in figure 15. Cantor constructed a new number by modifying the diagonal elements. He then assembled them into a new real number (*r*). One such digit-modifying scheme is as follows: if the digit is a 1, change it to a 2; if it's not a 1, change it to a 1. The number so formed from the diagonal elements 372211... is *r* = 111122.... This real number (*r*) is different in at least one digit from any of the real numbers in the array.

Since it was assumed that all real numbers were listed in the array, we have a contradiction to the assumption that the real numbers have cardinal number N_0; i.e., it can be placed in a one-to-one correspondence with the natural numbers. As shown above, the real numbers do not have the same cardinal number as the natural, rational, and algebraic numbers. Hence, there must exist number sets, other than *N*, *F,* and *A*, in the unit interval. We determine what they are in the next section.

3	1	7	5	6	2	.....
9	7	3	2	6	8	.....
5	3	2	7	4	9	.....
2	8	6	2	4	7	.....
3	7	9	5	1	3	.....
6	4	9	3	7	1	.....
.....	.....	.....	.....	.....	.....	.....
.....	.....	.....	.....	.....	.....	...

Figure 26. Cantor's diagonal process

Irrational (*I*) and Transcendental Numbers (*T*)

So far, we have the unit interval of real decimals packed with rational and algebraic numbers. Can we pack in more? We surely can. As indicated in figure 27, we see that the diagonal of a unit square has a length of $\sqrt{2}$. If we placed the diagonal on the *x*-axis, the ends of the diagonal lie on the points labeled 0 and $\sqrt{2}$. In addition, we know that $\sqrt{2}$ is not a rational number.

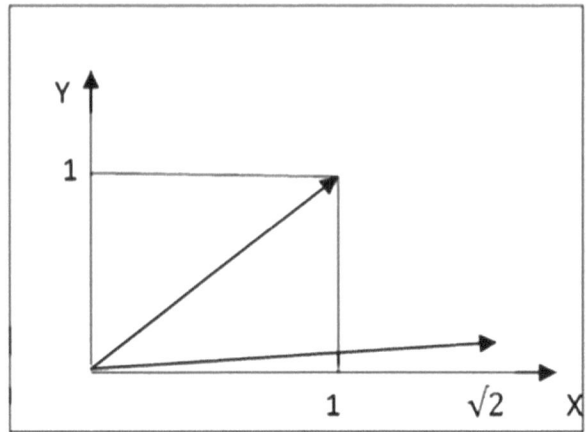

Figure 27. Diagonal of square

For those interested, the proof that √2 is not a rational number is given in appendix 4. Numbers that cannot be written as a ratio (p/q) are called irrational numbers. The irrational numbers are those numbers remaining when the rational numbers are removed from the real numbers (in set notation, $I = R - F$). The decimal part of an irrational number never ends or repeats. Some examples follow:

$$\sqrt{2} = 1.414213562...$$
$$\text{Pi } (\Pi) = 3.14159265...$$

In appendix 5, for those interested, we explain graphically how the continued application of the Cantor diagonal procedure to the arrays of rational number generates an unending sequence of irrational numbers (I). If you apply the Cantor diagonal procedure to the arrays of algebraic numbers instead of the rational numbers, you generate an unending stream of transcendental numbers (T). In set notation, $I = R - F$, and $T = R - A$. In both cases, a set with cardinal number N_0 is removed from a set with cardinal number (c). It can be shown that the remaining sets (I and T) retain the cardinal number of the original set. Thus, both I and T have cardinal number (c). The two Venn diagrams in figure 28 illustrate the relationships existing among the seven sets of numbers identified above.

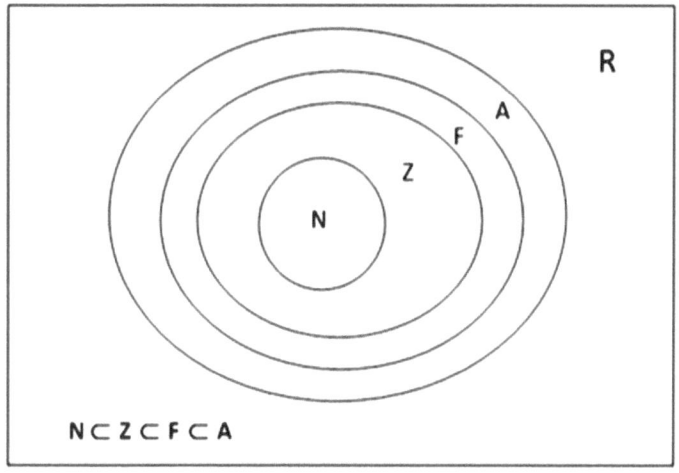

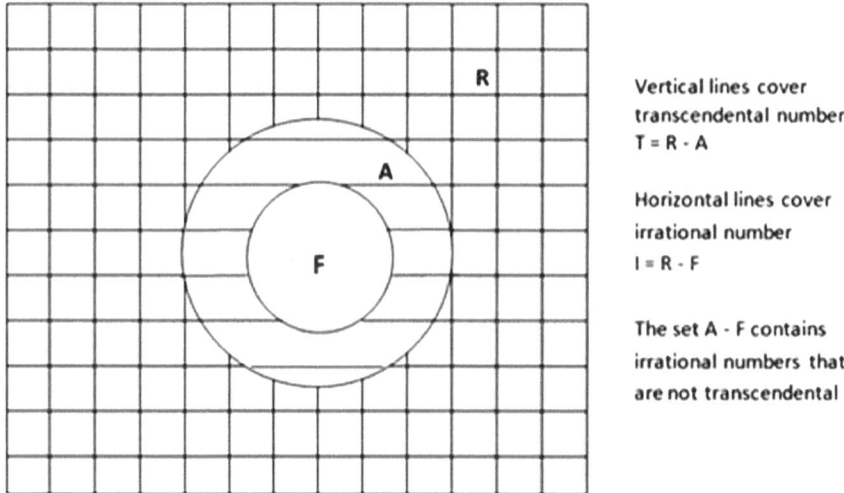

Figure 28. Subsets of *R* with cardinal number N_0 and c

Now that we have the real numbers completely filling the line segment, what can we say about the cardinality of the set of points in the Cartesian plane or even in a unit square? We would think that infinity times infinity would be larger than just infinity. Two dimensions surely have more points than a single dimension. Dealing with infinite sets is where our intuition goes astray. As shown in appendix 6, it is possible to place the points in the plane in a one-to-one correspondence with the points on a line. Thus, a two-dimensional square has the same cardinal number as a one-dimensional line. Further, n-dimensional space has the same cardinality as a one-dimensional space has.

Other than the structured sets of numbers defined above, there are many additional sets or classes of numbers that are defined by equations, processes, or algorithms. Of special interest is the set known as prime numbers.

Prime Numbers

Write down a sequence of consecutive integers starting with 2. Include at least fifty numbers in the sequence. Starting with 2, remove every number spaced two counts apart (divisible by 2). Then, from 3 on, remove every number divisible by 3. Number 4 will have been removed, so use 5 to remove all numbers spaced five counts apart. Continue the removal process for each remaining integer in your sequence of consecutive integers. If you continue this process correctly, the numbers remaining will be prime numbers, which are defined as numbers other than 1 and itself that can't be written as the product of two or more factors other than 1. As you go through the elimination process, you should note two properties. First, you only need to use the primes in the elimination process. Second, if there are N numbers in the sequence or array, you only have to use the primes up to the square root of N.

The process can be performed using a square array rather than a sequence of numbers. Such an array is shown in figure 29.

	2	3		5		7			
11		13				17		19	
		23						29	
31						37			
41		43				47			
		53						59	
61						67			
71		73						79	
		83						89	
						97			

Figure 29. Eratosthenes's sieve

This process of elimination was discovered by Eratosthenes in AD 100. The numbers removed in the process are called composite numbers. For example, $8 = 2 \times 2 \times 2 = 2 \times 4$, where 2 and 4 are factors of 8; and $15 = 3 \times 5$, where 3 and 5 are factors of 15. It should be noted that every composite number can be written as the product of primes. For example,

$15 = 3 \times 5; 22 = 2 \times 11; 24 = 2 \times 2 \times 2 \times 3; 21 = 3 \times 7; 30 = 2 \times 3 \times 5.$

Examples of primes less than 100 are as follows:

2, 3, 5, 7, 11, 13, 17, 19, 23, 29, 31, 37, 41, 43, 47, 53, 59, 61, 67, 71, 73, 79, 83, 97

As of April 2014, the largest known prime has 17,425,170 digits. Also, the separation between two consecutive primes cannot be larger than 70,000,000. The set of prime numbers (P) is an example of a proper subset of natural numbers. As indicated below, Euclid proved that the set (P) of prime numbers has the same cardinal number N_0 as the natural numbers.

Euclid (circa 300 BC) used a unique form of proof to demonstrate that there is an infinite number of primes. The proof is worthy of inclusion here because it shows that math works in a simple and elegant manner. Euclid formed the product of a random set of different primes:

$$P_1, P_2, P_3, P_4, P_5 \ldots P_n$$

He then proved that regardless of the primes in the product, it is always possible to find a prime that is not in the set. He formed the number (N).

$$N = P_1 P_2 P_3 P_4 P_5 \ldots P_n + 1$$

He noted that only two cases exist. Case 1: N is a prime. Case 2: N is not a prime. Case 1: if N is a prime, then N is a new prime number different from any of the P_i. Case 2: if N is not a prime, it is composite. Hence, there exists a prime number (P) that divides N. The prime (P) cannot be any of the P_i that form N. For if P is any of the P_i, then dividing N by P leaves a remainder of 1. Therefore, P is a prime different from any of the P_i. In both cases, a new prime number was found. Hence, for any original set of prime numbers, it is possible to enlarge the set with new and different prime numbers. Hence, the set of prime numbers is infinite.

NUMBER SYSTEMS

IN THE ABOVE paragraphs, we have described the structure of the real number continuum. In the description of that structure, we saw how mathematics used sets and logic to partition seemingly unrelated information into defined categories. Peano established a basis for the natural numbers. Starting with the set of natural numbers, other sets of numbers were defined using arithmetic or set operations. For example, the number 0 and the negative of the natural numbers were added to the natural numbers to form the integers. The set of rational numbers was defined as the numbers resulting from the ratio of two integers. The set of algebraic numbers was defined as a subset of the roots (solutions) to the general polynomial equation. Cantor used the diagonal process to prove that numbers other than the rational numbers existed. The sets of irrational numbers and transcendental were defined in the set of real numbers as set complements to the rational numbers and the algebraic numbers respectively.

What we haven't done is show how the natural numbers form composite numbers consisting of a whole part and decimal part separated by the decimal point, as shown below:

$$b_n\, b_{n-1}\, b_{b-2}\ldots b_0.\, a_1\, a_2\, a_3\ldots a_0\ldots$$

Decimal Numbers

Composite numbers consist of a sequence of digits. If the digits range from 0 to n, the number system is said to be a base $n + 1$ system. The two number systems with the greatest utility are the base 10 and the base 2 systems. In the base 10 system, the digits can assume values from 0 to 9. In the decimal part, the a_i forms a sequence of digits that may terminate or continue indefinitely. The whole part is always an integer. Such a system is called the decimal number system. The descriptor "decimal" is dropped when it is understood that the number system in use is the decimal (base 10) system. When there is doubt, the number will have a subscript indicating its base number. For example, the subscript 5 following the number 324.132_5 indicates the number is a base 5 number. In the base 10 system, the position of a digit in the number determines the power of 10 to which that digit is multiplied. For example, in the number 5234,

Position	3	2	1	0
Power	10^3	10^2	10^1	10^0
Number	5	2	3	4
Value	5 × 1000	+ 2 × 100 +	3 × 10 + 1 × 4 = 5234.	

Positions are numbered from the left of the decimal point and increase from 0. For example, in the number 5234, the integer 5 in the leftmost position has a higher value or significance than the digit 2 to the right of it. The position value of the 5 is expressed in thousands 10^3, the 2 in hundreds 10^2, the 3 in tens 10^1, and the 4 in units 10^0. Arithmetically, we express the total as a sum: $5 \times 10^3 + 2 \times 10^2 + 3 \times 10 + 4 \times 1 = 5234$.

Each position in a number is assigned a base value that is a power of ten. Starting from the decimal point and proceeding to the left, the powers of 10 increase from 0 to n. Starting from the decimal point and moving to the right, the powers of 10 increase negatively from –1 to –n. For example,

n......	6	5	4	3	2	1	0 .	–1	–2	–3	–4
10^n.....	10^6	10^5	10^4	10^3	10^2	10^1	10^0	10^{-1}	10^{-2}	10^{-3}	10^{-4}

where

10^{11} = 100 billion = 100,000,000,000 10^{-11} = 0.00000000001
10^{10} = 10 billion = 10,000,000,000 10^{-10} = 0.0000000001
10^9 = 1 billion = 1,000,000,000 10^{-9} = 0.000000001
10^8 = 100 million = 100,000,000 10^{-8} = 0.00000001
10^7 = 10 million = 10,000,000 10^{-7} = 0.0000001

10^6 = 1 million = 1,000,000
10^5 = 1 hundred thousand = 100,000
10^4 = ten thousand = 10,000
10^3 = one thousand = 1,000
10^2 = one hundred = 100
10^1 = ten = 10
10^0 = one = 1

Consider the number 3,160,402.135. The number 3,160,402 to the left of the decimal point is called the whole or integral part. The number 135 is called the decimal or fractional part. The value of the number is the sum of the products of the individual digits and 10 to the power of their position in the number.

$3 \times 10^6 + 1 \times 10^5 + 6 \times 10^4 + 4 \times 10^2 + 2 \times 10^0 + 1 \times 10^{-1} + 3 \times 10^{-2} + 5 \times 10^{-3}$ = 3,160,402.135

This scheme of writing numbers was developed over many years. It is now in standard use throughout most of the world. There is a reason for this standard to be in general usage. Humans started tallying or counting with their available body parts. It was natural for the ten fingers to serve as a tallying device when counting a small number of objects. It was a natural extension to use multiples of ten when a larger number of objects had to be counted.

However, while we were all educated to write numbers in base 10, numbers can be written using a base other than 10. Included in the history of numbers are civilizations using number systems having a base other than 10. Mesopotamians, circa 3400 BC, used a base 60 system. Mexico's Mayans used a combination of base 4 and base 5 to form a base 20 system. Mathematicians always had an interest in studying number systems that used a base other than base 10. The one that found the most use was the number system that used the digit 2 as its base. That number system is called the binary number system. Computations in modern-day digital computers are performed using base 2 arithmetic. A description of the base 2 number system follows.

Binary Numbers

Binary numbers are written using a sequence of digits that are 0s and 1s instead of the digits 0 through 9. In the binary system, we only use the digits 0 and 1. The numbers 0 and 1 are called binary digits. In practice, the name is contracted to bits (binary digits). With decimal numbers, we use powers of 10 as multipliers to determine the value assigned to each position in a number. With binary numbers, we use powers of 2 to determine the value assigned to each position in a number. For example, the number 51 in base 10 is written 110011 in base 2. For example, in the number 110011:

Position	5	4	3	2	1	0
Power	2^5	2^4	2^3	2^2	2^1	2^0
Number	1	1	0	0	1	1
Value	$1 \times 32 +$	$1 \times 16 +$	$0 \times 8 +$	$0 \times 4 +$	$1 \times 2 +$	$1 \times 1 = 51_{10}$

Obviously, the length of a number written in binary (base 2) is much longer than the same number written in decimal (base 10). Employing the base 2 representation for numbers has many advantages over the base 10 representation. First, base 2 numbers are easily represented by electronic circuits. Only two levels are required. On = 1, and off = 0. The binary system is used in designing the arithmetic and information processes performed by digital computers. Second, arithmetic is much simpler when performed in binary than in decimal. The rules for binary arithmetic follow.

Addition in binary: $1 + 0 = 0 + 1 = 1$

$1 + 1 = 0$ carry 1

Multiplication in binary: $1 \times 0 = 0 \times 1 = 0$

$1 \times 1 = 1$

Subtraction: 1 − 0 = 1

1 − 1 = 0
0 − 1 = 0 borrow 1

Division is a series of comparisons, multiplications, and subtractions. As in a base 10 division, a base 2 division may end or result in a remainder. Examples of each binary operation follow.

Binary addition: We convert 21 + 13 = 34 to base 2 and add. The first row of 1s contains the carries for the addition.

$$\begin{array}{r} 1\,1\,1\,1 \\ 21_{10} = 1\,0\,1\,0\,1 \\ 13_{10} = 1\,1\,0\,1 \\ \hline 1\,0\,0\,0\,1\,0 \end{array}$$

Binary subtraction: 21 − 13 = 8. The first row of 1s are the borrows.

$$\begin{array}{r} 1 \\ 1\,0\,1\,0\,1 \\ 1\,1\,0\,1 \\ \hline 0\,1\,0\,0\,0 \end{array}$$

Binary division: 45 divided by 6 = 7.5

$$\begin{array}{r} 1\,1\,1\,.\,1 \\ 1\,1\,0\,\overline{)1\,0\,1\,1\,0\,1} \\ 1\,1\,0 \\ \hline 1\,0\,1\,0 \\ 1\,1\,0 \\ \hline 1\,0\,0\,1 \\ 1\,1\,0 \\ \hline 0\,1\,1\,0 \end{array}$$

Binary multiplication: 42 × 13 = 546

```
     1 0 1 0 1 0
         1 1 0 1
     1 0 1 0 1 0
   1 0 1 0 1 0 0
   1 0 1 0 1 0
 1 0 0 0 1 0 0 0 1 0
```

Converting Base 2 Numbers to Base 10

To facilitate the conversion process, we normally prepare a table containing the value of each digit according to its position in the base 2 number. We count positions starting at the decimal point. The exponents of 2 start at 0 and increase positively from the left of the decimal point for the whole part of the number. In the decimal part of the number, exponents start at –1 and increase negatively from the right of the decimal point. The value of each position in a binary number is illustrated below:

$$\ldots 2^7\ 2^6\ 2^5\ 2^4\ 2^3\ 2^2\ 2^1\ 2^0\ .\ 2^{-1}\ 2^{-2}\ 2^{-3}\ 2^{-4}\ 2^{-5}\ 2^{-6}\ 2^{-7}\ldots$$

where, $2^7 = 128$ $2^{-7} = 0.0078125$
$2^6 = 64$ $2^{-6} = 0.015625$
$2^4 = 16$ $2^{-5} = 0.03125$
$2^3 = 8$ $2^{-4} = 0.0625$
$2^2 = 4$ $2^{-3} = 0.125$
$2^1 = 2$ $2^{-2} = 0.250$
$2^0 = 1$ $2^{-1} = 0.500$

In the conversion of a binary number to its decimal equivalent, the position of each 1 in the binary number determines the amount it contributes to the base 10 equivalent. For example, in the binary number 1 0 0 1, there are 1s in the 0 and 3 positions. The 1 in the 0 position contributes $2^0 = 1$. The 1 in the third position contributes $2^3 = 8$. The base 10 number is the sum of the contributions from the 1s in the binary number. In this example, binary 1 0 0 1 converts to decimal 9. For each 1 in the binary number, you determine its position and assign its base 10 value according to the listing above. The base 10 equivalent is the sum of the base 10 values for the individual bits in the base 2 number.

The binary to base 10 procedures are best illustrated through examples. In the first example, we shall convert

1 0 1 0 1 1 0 1 . 1 1 0 1 into its base 10 equivalent, 173 .8125.

In the conversion, we start counting from the decimal point and note the positions of the 1s in the binary number. For the integral part of the number, the positions are 0, 2, 3, 5, and 7. These positions have base 10 values $2^0 = 1$, $2^2 = 4$, $2^3 = 8$, $2^5 = 32$, and $2^7 = 128$. Together they equal to 173.

$$
\begin{array}{ccccccccc}
1 & 0 & 1 & 0 & 1 & 1 & 0 & 1 & \\
128 + & 0 + & 32 + & 0 + & 8 + & 4 + & 0 + & 1 & = 173 \\
1 \times 2^7 + & 0 \times 2^6 + & 1 \times 2^5 + & 0 \times 2^4 + & 1 \times 2^3 + & 1 \times 2^2 + & 0 \times 2^1 + & 1 \times 2^0 &
\end{array}
$$

In converting the decimal part (.8125) of the number, we use the same procedure as above. However, we use negative powers of 2.

$$
\begin{array}{cccc}
.1 & 1 & 0 & 1 \\
1 \times 2^{-1} + & 1 \times 2^{-2} + & 0 \times 2^{-3} + & 1 \times 2^{-4} \\
0.500 + & 0.250 + & 0 + & 0.0625 = 0.8125
\end{array}
$$

The result is 10101101.1101 = 173.8125.

Converting Base 10 Numbers to Base 2

In the above example, we easily converted a base 2 number into its equivalent base 10 number. However, converting a base 10 number into its base 2 equivalent is not as straightforward. The numbers to be converted consist of an integral part and a decimal part. For example, 5 8 2 . 3 6 1 and 1 1 0 1 . 0 1 0 1. The base 10 to base 2 conversion procedures can be tedious and error-prone. However, there is assistance available. If you search the Internet, you will find ready-to-use decimal-to-binary conversion programs. You can use them to check the accuracy of any conversion you performed manually. However, to really get to understand how numbers, using different bases, are structured and related, you need to go through the steps in the conversion processes described below.

All base 10 to base 2 conversion processes (algorithms) perform operations on the base 10 number that show the positions of the 1s in the base 2 equivalent. In the example above, the binary number 1 0 1 0 1 1 0 1 . 1 1 0 1 had 173.8125 as its base 10 equivalent. The 1s in the binary number represented powers of 2 that were summed to obtain the value of the base 10 number. In converting the base 10 number to base 2, we have to reverse the process. We process the base 10 number to locate the positions of the 1s and 0s in the binary number. How this is accomplished is best explained through examples.

The conversions are performed separately on the whole and decimal parts in order to select the algorithm best suited to convert each of the two parts. Having a table containing the values for the powers of 2 will expedite the conversion process. A partial table is shown below:

$2^7 = 128$ $2^{-7} = 0.0078125$
$2^6 = 64$ $2^{-6} = 0.015625$
$2^4 = 16$ $2^{-5} = 0.03125$
$2^3 = 8$ $2^{-4} = 0.0625$
$2^2 = 4$ $2^{-3} = 0.125$
$2^1 = 2$ $2^{-2} = 0.250$
$2^0 = 1$ $2^{-1} = 0.500$.

Highest Power Division Algorithm

In this example, we convert the base 10 number 3625 to a number written in base 2. We know that the conversion of the base 10 number 3625 to base 2 will result in a sequence of binary numbers (1s and 0s) whose positions have values equal to the powers of 2 for that position. In this example, the powers of 2 for that sequence are given below:

$$2^{11}, 2^{10}, 2^9, 2^8, 2^7, 2^6, 2^5, 2^4, 2^3, 2^2, 2^1, 2^0$$

The binary equivalent to 3625 will be a sequence of 1s and 0s. Each 1 in the sequence will contribute 2 to the power of its position in the sequence. Hence, we seek the power of 2 that yields the largest number that divides into 3625. The largest power of 2 (2^{11} = 2048) that divides 3625 is 11. In the binary number, a 1 is placed in the twelfth position of the binary number.100000000000 (note that the position numbers start at 0). To be subtracted is 2048 from the base 10 number 3624 with a remainder of 1577. We now have to locate the positions of the 1s in the binary number that contribute to the remainder 1577. We repeat the process on this remainder. We reduce the power of 2 by 1 position and check to see if 2^{10} =1024 divides into the remainder. If it is a divisor, that position in the binary number receives a 1; if not, a 0.

At this point in the process, we have located two positions contributing a total of 2048 + 1024 = 3072 to the total 3625, leaving a remainder 553. We continue the process of seeking powers of 2 that divide into the remainder. We find that 2^9 = 512 divides 553, leaving a remainder 41. A 1 is placed in the ninth position. We operate on the remainder until we arrive at the power of 2 being 1 or 0 for even and odd base 10 numbers respectively. The result of the conversion process follows.

3625,	1577,	553,	41,	41,	41,	41,	9,	9,	1,	1,	1
2048,	1024,	512,	256,	128,	64,	32,	16,	8,	4,	2,	1
1577,	553,	41,				9		1			1
1	1	1	0	0	0	1	0	1	0	0	1
2048 +	1024 +	512 +				32 +		8 +			1 = 3625

Division by 2 Algorithm

In this example, we will use a different algorithm to convert the number 3625 to its binary equivalent. In the algorithm, we first note whether the number to be converted is odd or even. If it is odd, the binary number will also be odd. Hence, the least significant bit of the binary number will be a 1. We subtract 1 from the number and divide the remainder by 2. In dividing by 2, we shift the binary number one significant place to the right. That operation exposes the next least significant bit in the binary number. Again, if the resulting number is odd, the next least significant bit will be a 1. If the number is even, the bit will be a 0. We continue dividing the remainders by 2 and record the resulting bits and their positions in the binary number. Commas separate each step in the conversion process, as shown below:

3625, 3624/2, 1812/2, 906/2, 453, 452/2, 226/2, 113, 112/2
　　1　　　　0　　　0　　1　　　　0　　1

56/2, 28/2, 14/2, 7, 6/2, 3, 2/2, 1
　0　　0　　0　　1　　1　　1

Since the bits were exposed starting with the least significant bit, we have to read the bits in the reverse order. Reading the bits in reverse, the binary number is

1 1 1 0 0 0 1 0 1 0 0 1.

Summing the powers of 2 represented by the positions of the 1s,

2048 + 1024 + 512 + 32 + 8 + 1 = 3625

CONVERTING DECIMAL NUMBERS

WHEN CONVERTING WHOLE numbers (integers), the conversion process may be long, but it will always terminate. Such is not the case when converting decimal (fractional) numbers. As stated in the text, decimal numbers assume several forms. They may terminate, continue to expand in a cyclical pattern, or continue indefinitely. In a rational decimal, the digits either terminate or continue in cycles. For example, $1/8 = 0.125$, $3/11 = 0.272727...$ In irrational decimal numbers, their digits are noncyclical and continue indefinitely: $\sqrt{2} = 1.414213562...$ When a base 10 decimal is converted to a base 2 decimal, the pattern of bits in the base 2 equivalent will exhibit the same pattern as the base 10 number.

When we convert base 10 decimal numbers, we deal with numbers that are less than 1. We have to start from the right of the decimal point and find the positions in the binary number that require a 1. Since the conversion may be cyclical or not terminate, we have to be able to handle large negative powers of 2. This leads to numbers like $2^{-10} = 0.0009765625$ compared to $2^{10} = 1024$.

Highest Power Division Algorithm

As an example of the conversion process, let us start with the base 10 decimal 0.695. We use the highest power of 2 division algorithm to accomplish the conversion. In the conversion process, we have to find the largest number expressed as a negative power of 2 that divides into the number 0.695. We find that -1 yields $2^{-1} = 0.500$. Note also that the position -1 starts to the right of the decimal point. Larger negative powers of 2 locate positions farther to the right. In the example, we list the entries in a column rather than a row because of the length of the entries.

0.695
<u>0.500</u> –1
0.195
0.250 0
<u>0.125</u> –3
0.0700
<u>0.0625</u> – 4
0.0075
0.031250 0
0.015625 0
0.00781250 0
<u>0.00390625</u> –8
0.003593750
0.0019531250 –9
0.0016406250
0.0009765625 –10
0.00066416250
0.00048828125 –11
0.00017598125
0.000244140625 0
0.0001220703125 –13
……………………
……………………
……………………

Continuing this process, we obtain the binary number below. We use parentheses to indicate the cycles in the expansion. We stop the process once we recognize a cyclical pattern.

101 {10001110101110000101} {10001110101011110000101}…

It is clear that if we continued the process, we would be handling negative powers of 2, which result in a very large number of digits.

Division by 2 Algorithm

To compare the largest power of 2 algorithm to the division by 2 algorithm, we convert 0.695 to its binary equivalent using the latter algorithm. Since we are dealing with a decimal, the first position to the right of the decimal point will have value $2^{-1}= 0.5$. Hence, we will be dividing with negative powers of 2, or equivalently, multiplying by 2. Multiplying by 2 shifts the number one position to the left. If the shift results in a number greater than 1, a 1 is placed in the binary number. If the result is a number less than 1, a 0 is placed in the binary number. Only a part of the process is completed below:

.695 × 2 = 1.39 - 1
.390 × 2 = .780 - 0
.780 × 2 = 1.56 - 1
.560 × 2 = 1.12 - 1
.120 × 2 = .240 - 0
.240 × 2 = .480 - 0
.480 × 2 = .960 - 0
.960 × 2 = 1.920 1
.920 × 2 = 1.84 - 1
.840 × 2 = 1.680 - 1
.680 × 2 = 1.340 - 1
.340 × 2 = .680 - 0
.680 × 2 = 1.360 - 1
. . . .

Comparing the two conversion processes, it is obvious that the division by 2 algorithm results in the handling of much smaller numbers. Hence, the division by 2 algorithm is the better choice.

CARDINAL NUMBERS

IN OUR JOURNEY, along the landscape of the real number continuum, we encountered seven sets of numbers: natural (N), integers (Z), rational (F), algebraic (A), transcendental (T), irrational (I), and real (R). The seven sets of numbers are all infinite. We used Venn diagrams (figures 14, 17, and 19) to show how the seven sets are related not by size but by inclusion. We understand the terms "size," "larger," or "smaller" when used to compare finite sets. A set containing twelve elements is larger than a set containing six elements. The use of these terms is confusing when used to compare infinite sets. For example, if we compare the set of rational numbers (F) with the set of natural numbers (N), we readily note that F has more elements than N. Would you say that the size of F is larger than the size of N? We can say that N is a proper subset of F because F contains N as a subset and F has numbers that are not in N. However, F and N have the same cardinal number; i.e., the elements of F can be placed in a one-to-one correspondence with the element of N. In that sense, the two sets have the same number of elements.

To reduce the confusion when comparing infinite sets, we employ Cantor's definition, which states that "if there exists a one-to-one correspondence between the elements of two sets (S and T), the sets are said to have the same cardinal number." Sets satisfying the one-to-one correspondence criteria are also said to be in the same equivalence class. Thus, for infinite sets, their cardinal number is an intuitive measure of their size. Every set has a cardinal number. To denote the cardinal number for a set (A), we place a bar over the symbol A. For example, to denote that two sets, A and B, have the same cardinal number, we place a bar over A and B and write

$$\bar{A} = \bar{B}.$$

The bar over the set symbol is not to be confused with the complement of the set. The first four sets of numbers—the natural numbers (N), the integers (Z), the rational numbers (F), and the algebraic numbers (A)—were all determined to have the same cardinal number N_0 (aleph null). They belong to the equivalence class N_0 because their elements can be counted off, that is, paired in a one-to-one correspondence with the natural numbers (N). Cardinal numbers establish an order, denoted by the symbol <, among equivalence classes. The order starts with the natural numbers and continues through classes of infinities. The natural numbers are called finite cardinal numbers. The cardinal numbers for infinite sets are called transfinite cardinal numbers.

$$1 < 2 < 3 < 4 ... < N_0 < c < f ...$$

For finite sets, each natural number determines an equivalence class. For example, the cardinal number 7 establishes an equivalence class of sets, with each set having seven elements. Every set having seven elements is a member of the same equivalence class. Let us review how the four equivalence classes were defined or constructed in the text above. First, we stated that a set is finite if its elements can be placed in one-to-one correspondence with a set of (n) natural numbers. Excluding the empty set, a set was deemed to be infinite if it was not finite. A set (S) was also said to be infinite if it had a proper subset (S_j) whose elements could be placed in a one-to-one correspondence with the elements of S. The natural numbers were identified as our first infinite set of numbers belonging to the equivalence class N_0. We showed how the integers, rational numbers, and the algebraic numbers were placed in a one-to-one correspondence with the natural numbers. They were placed in the same equivalence class as the natural numbers. We used the Cantor diagonal procedure to show that numbers other than rational and algebraic existed. The irrational numbers were defined as the set of real numbers complementary to the rational numbers. The transcendental numbers were identified as the set of real numbers complementary to the algebraic numbers. The real numbers (R), the irrational numbers (I), and the transcendental numbers (T) were found to have cardinal number (c).

Using the one-to-one correspondence, we were able to decompose the subsets of real numbers into two equivalence classes, R_1 and R_2. The class R_1 has N, Z, F, and A as members with cardinal number N_0. The equivalence class R_2 has *I*, *T*, and *R* as members with cardinal number (*c*). Many set theorists question, "Are there cardinal numbers between N_0 and *c*?" They agree that the answer to that question is undecided. Depending on how set theorists defined their sets, justifiable answers come back: yes and no. However, there is agreement among set theorists that N_0 is the smallest cardinal number.

Now the question is "Do infinite sets having cardinal numbers other than N_0 and *c* exist?" To answer that question, we have to define what we mean by the symbols = and < when used with respect to cardinal numbers reepresenting different equivalence classes. We know what the symbols mean when used with real numbers.

When dealing with infinite sets, the symbols = and < take on a new meaning. When we say that two sets—*A* with cardinal number *a* and *B* with cardinal number *b*—have the same cardinal number, we write *a* = *b*. Saying *a* = *b* is equivalent is equivalent to saying that *A* has a subset that can be placed in a one-to-one correspondence with *B* and vice versa. However, if the sets *A* and *B* have different cardinal number *a* and *b*, then either *a* < *b* or *b* < *a*. Saying that a < b is equivalent to saying that *B* has a subset that can be placed in a one-to-one correspondence with *A* but not vice versa. The symbol < establishes an order among the cardinal numbers very much in the way the symbol < orders the natural numbers. The cardinal numbers 3 and 7 are related as in 3 < 7 since a subset of any 3 elements from a set of 7 elements can be placed in a one-to-one correspondence with the set of 3 elements. The reverse is not true.

Now we have to answer the question "Relative to the order <, are there sets with cardinal numbers beyond *c*?" To answer that question, we have to introduce a new set called the power set.

Power Set

From any set (*A*) having a finite number of elements, it is always possible to construct proper subsets of *A* containing some of its elements. The subsets can be formed by assembling the elements one at a time, two at a time, etc. Associated with the set (*A*) is the set A^*, whose elements are all the subsets of *A*. A^* is called the power set of *A*. If the set (*A*) has *n* elements, its power set A^* will have 2^n subsets as elements. As examples of power sets, we construct the power sets for four sets consisting of 1,

2, 3, and 4 elements. The elements of the sets could be labeled *a*, *b*, *c*, and *d*. However, since the size of the set—i.e., the number of elements in the set is also its cardinal number (*CN*)—we will use the natural numbers as labels for the elements of the sets.

For *CN* 1, the power set $1^* = [\{0\}, \{1\}]$ and $1 < 2^1 = 2$.

For CN 2, the power $2^* = \{\{0\}, \{1\}, \{2\}, \{1, 2\}\}$, and $2 < 2^2 = 4$.

For CN 3, the power set $3^* = \{\{0\}, \{1\}, \{2\}, \{3\}, \{1, 2\}, \{1, 3\}, \{2, 3\}, \{1, 2, 3\}\}$, and $3 < 2^3 = 8$

For CN 4, the power set $4^* = [\{\{0\}, \{1\}, \{2\}, \{3\}, \{4\}, \{1, 2\}, \{1, 3\}, \{1, 4\}, \{2, 3\}, \{2, 4\}, \{3, 4\} \{1, 2, 3\}, \{1, 2, 4\}, \{1, 3, 4\}, \{2, 3, 4\}, \{1, 2, 3, 4\}\}]$, and $4 < 2^4 = 16$.

The examples illustrate the relationship existing between the size of a finite set and its power set. As the size of the finite set increases linearly, the size of its power set increases exponentially. This relationship between the number of elements in a set and the number of its subsets remains true for any set (*A*) containing a finite number of elements; i.e., if the set (*A*) contains *n* elements, its power set (*A**) contains 2^n subsets. We readily deduce that for *n* finite,

$$n < 2^n.$$

Does a similar relationship hold true between a set (*A*) and its power set (*A**) when the set (*A*) contains an infinite number of elements? To answer that question, we have to accept several mathematical facts. First, it can be stated that any set (*A*) does not have the same cardinal number as its power set (*A**). Using the set *N* of natural numbers as an example, where *N* has cardinal number N_0, it can be stated that its power set (*N**) has cardinal number 2^{N_0} and that

$$N_0 < 2^{N_0} = c$$

We can make the same statement for the set of real numbers (*R*), where *R* is the set of all finite and infinite decimal fractions in the real number interval between two successive natural numbers. The cardinal number of R is c. The power set R^*, the set of all subsets of R, does not have the same cardinal number as R has. If we label f as the cardinal number of R^*, we can state that $c < 2^c = f$. For real numbers, we now have an order for transfinite cardinal numbers:

$$N_0 < c < f$$

Can we continue the order (<) to include an infinite sequence of transfinite cardinal numbers?

Exponentiation

Exponentiation of a cardinal number is the process of using the cardinal number as an exponent in a mathematical equation. As used herein, the cardinal numbers are used as exponents of 2. As noted in the previous section, $2^{N_0} = c$. We know that N_0 is the first cardinal number. And we also know that for any cardinal number (p),

$$P < 2^p.$$

We can view exponentiation as a successor function for the cardinal numbers. Applying the successor function to N_0, we obtain its successor cardinal number, $2^{N_0} = c$. The next cardinal number is $2^c = f$. When the process is continued for the set of the real numbers, we have

$$N_0 < c < f < 2^f.$$

We can use the process of exponentiation to write

$$f < 2^f = g < 2^g = h < 2^h = \ldots$$

This process yields an infinite sequence of transfinite cardinal numbers that can be ordered and labeled as

$$a_1 < a_2 < a_3 < \ldots < \ldots$$

Does this infinite sequence of cardinal numbers constitute the totality of cardinal numbers? The answer is no. We can define a set S consisting of sets A_i from each cardinal number equivalence class a_i to form the sum $S = \Sigma A_i$, $i = 1, 2, 3....$ For example, from the equivalence class $a_1 = N_0$, we could select the set of algebraic numbers A. From the equivalence class $a_2 = c$, we could select the set of irrational numbers I. If we continued the process for each equivalence class in the sequence, the set S will have cardinal number s, where $a_i < s$ for all i. The cardinal number (s) differs from and is of higher order than any cardinal number a_i in the sequence.

Such is the magic of numbers. The natural numbers increase unit by unit without end. The cardinal numbers are ordered through < without end. Peano's axioms order the natural numbers, and the process of exponentiation orders the cardinal numbers. In our ever-expanding universe, we experience numbers from 0 to the extremely large, but never infinite. Even when we divide by 0, that infinity is considered an undefinable singularity. M. C. Escher, in his Circle Limit series, tried to capture the notion of infinity. His objective was to portray the infinitely small, not the infinitely large. That is the best we can do on paper. However, through our minds and the use of logic, sets, and cardinal numbers, we are able to abstractly visualize infinite and beyond.

APPENDIX 1

Converting Cyclical Decimals to Rational Form (p/q)

FIRST, WE NOTE that cyclical decimals consist of sequences of digits of two types: the terminating sequence and the cyclical sequence. A decimal consisting solely of a terminating sequence can be considered a cyclical decimal if an unending sequence of zeros is appended to it. We will use three examples to illustrate the conversion procedures. Example 1 is a terminating sequence of three digits, example 2 consists of a cyclical sequence of three digits, and example 3 has two digits followed by a cyclical sequence of three digits. The number of digits in the cyclical sequence is called its period. In the examples, you should note that the conversion process eliminates the entire cyclical sequence to arrive at integral values. Examples illustrating the conversion processes follow.

Example 1: terminating decimal $x = 0.375$

When the decimal terminates after n digits, the first step is to multiply the decimal by 10^n. This step converts the decimal to a whole number. In this example, with $n = 3$, we obtain $10^3 \times 0.375 = 375$. After the multiplication, the decimal becomes the whole number (375). Step 2 is to divide the whole number by 10^n to change the whole number to a fraction. In this example, with $n = 3$, we obtain $375/1000$. When reduced to lowest terms, we get the fraction $3/8$.

Example 2: immediate start of period $x = 0.361361$

When the decimal is cyclical and has a period of n digits, the first step is to multiply both sides of the equation by 10^n. This converts the decimal to a whole number with a decimal part. In this example, the period $n = 3$. Multiplying x by 10^3 yields $1000x = 361.361361...$. Step 2 is to subtract the original equation from the latter equation to remove the decimal and obtain the fraction.

$1000x = 361.361361$
$x = .361361$
$999x = 361$
$x = 361/999$

Example 3: delayed start of period $x = 0.57361361$

When the delay in the decimal is n digits, step 1 is to multiply the both sides of the equation by 10^n. This changes the original number to a whole number plus a decimal. In this example, $n = 2$. When we multiply the equation by 10^2, the result is $100x = 57.361361...$. As in example 2, step 2 is to determine the period (p) of the repeating digits. In this example, the period (p) = 3. When we multiply the latter equation by 10^3, the result is $100000x = 57361.361361...$. Subtracting the equation from step 1 yields the result below:

$100000x = 57361.361361...$
$100x = 57.361361...$
$99900x = 57304$ or
$x = 57304/99900$
$x = 14326/24975$

APPENDIX 2

Rational Numbers Have Cardinal Number N_0

IN THIS APPENDIX, we place the rational numbers into a one-to-one correspondence with the natural numbers. To do so, we will use two simple equations, a little analytic geometry, and triangular numbers. First, let's have a description of triangular numbers. Triangular numbers are the solutions to the equation

$$T_n = n(n+1)/2 \text{ for } n = 1, 2, 3....$$

For example, when $n = 3$, we have $3 \cdot 4/2 = 6$; when $n = 4$, we have $4 \cdot 5/2 = 10$. In figure A1.1, you can see why these numbers are called triangular.

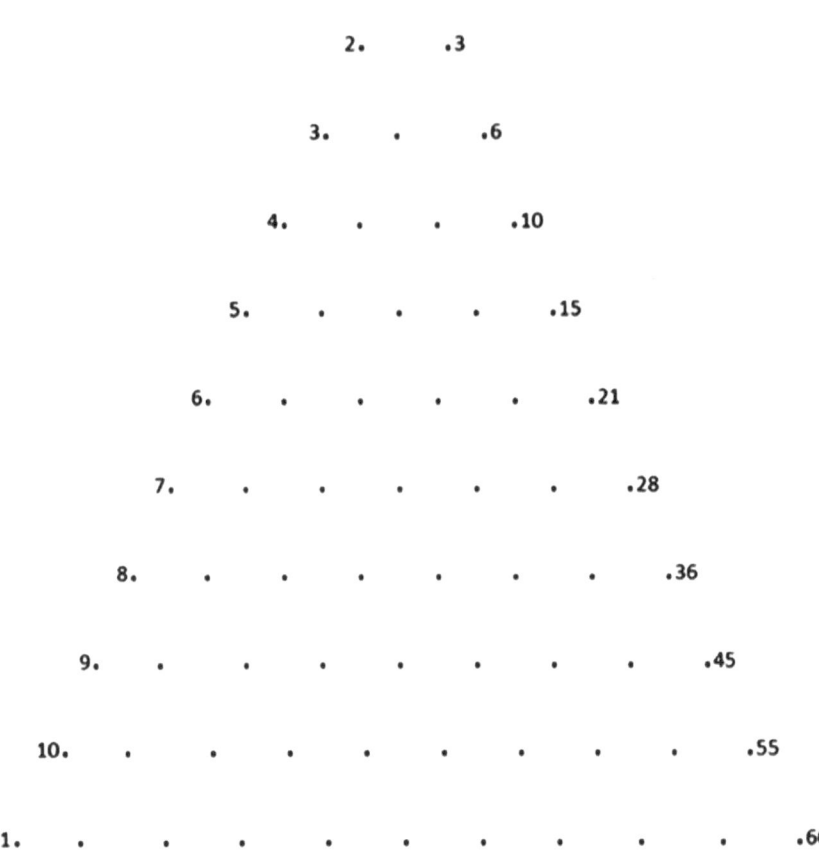

Figure A2.1. Triangular numbers

As you add the number of dots, row upon row, from top to bottom, you get the triangular numbers shown at the right of the triangle. In figure A2.2, we use analytic geometry to construct the lines for the equations $x + y = n$ and $n = 2, 3, 4...$ in the first quadrant of the x-y plane.

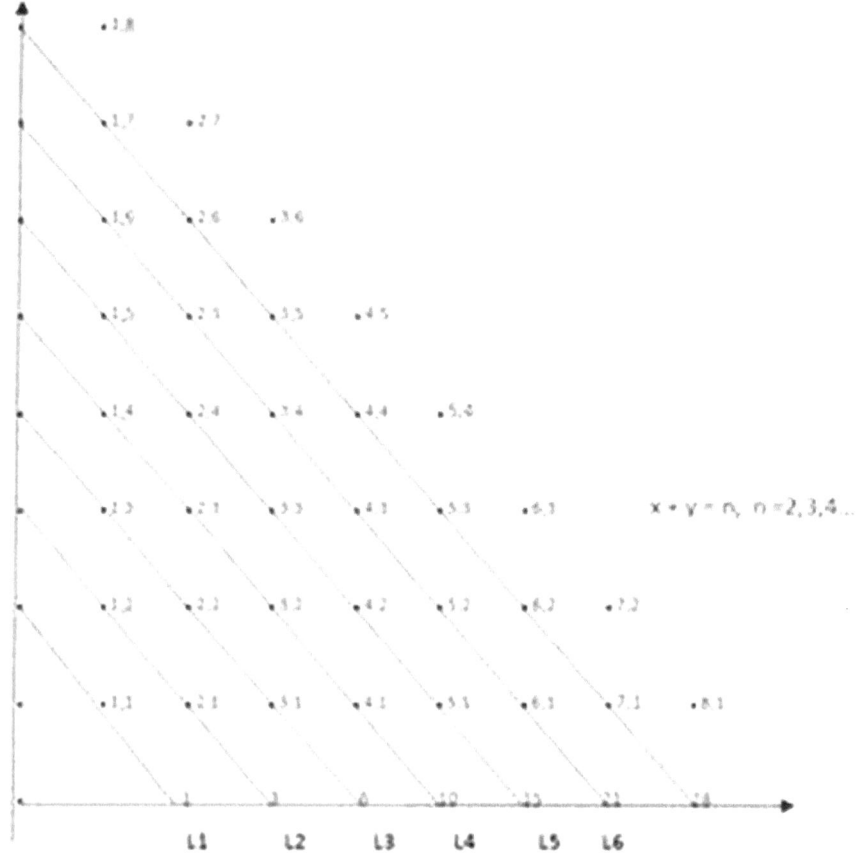

Figure A2.2. Plot of rational numbers

From bottom to top, we label the lines L_i, $i = 1, 2, 3, 4....$ On the lines, we note that the integral solutions to the equation $x + y = n$ (where $n = 2, 3, 4, 5, 6, 7...$) fall on dots representing the dot array for triangular numbers. The integral solutions for $x + y = n$ (where $n = 2$) yields $x = 1$ and $y = 1$ or $(1, 1)$ as a point on line L_1. Where $n = 3$, we obtain the solutions $(1, 2)$ and $(2, 1)$ as points on line L_2. Where $n = 4$, we obtain $(1, 3)$, $(2, 2)$ and $(3, 1)$ as points on line L_3. Each of these solutions represent a rational number since $x/y = p/q$. For example, $(1, 3)$ is the rational $1/3$. Hence, in the quadrant, we are able to list all the rational numbers p/q, having p and q as positive integers. It is interesting to note that the first row of rational numbers includes all the natural numbers.

Each additional row contains the natural numbers divided by the row number.

To construct the one-to-one correspondence, we start with line 1 and label the dots (rational numbers) from the y-axis to the x-axis with consecutive natural numbers. Using the notation (u, v), where u is the natural number and v is the rational number in the array, we arrive at the following sequence: $(1, 1)$, $(2, \frac{1}{2})$, $(3, \frac{2}{1})$, $(4, \frac{1}{3})$, $(5, \frac{2}{2})$, $(6, \frac{3}{1})$…. It is obvious that the pairings can be continued indefinitely to complete the one-to-one correspondence between the rational and natural numbers.

A question remains. Given a natural number, say, 14, is it possible to determine the rational number to which it was paired? The answer is yes. To find the rational number, we refer to figure A2.2. We note that the triangular numbers are written at the end of lines along the x-axis. The triangular numbers identify the rational numbers listed to that point. The triangular number at the end of line L_4 is 10. Since we are looking for the rational number at the natural number 14, that rational number will have to be on the next line (L_5). Counting from 10 at the top of line L_5, we advance four positions to the rational number 4/2 at position 14.

APPENDIX 3

Algebraic Numbers

IN THIS APPENDIX, we show how the algebraic numbers are placed in a one-to-one correspondence with the natural numbers. The set (A) of algebraic numbers is defined as the solutions (called roots) to the general algebraic equation,

$$a_0 x^n + a_1 x^{n-1} + a_2 x^{n-2} + \ldots + a_{n-1} x + a_n = 0,$$

where the degree of the equation (n) is a natural number, the a_i, i = 1, 2, 3... are integers, and $a_0 > 0$. The degree of the equation (the highest power of x), not the number of terms, determines the number of solutions (roots) to the equation. Depending on the values of the a_i, each equation of degree (n) may contain from 1 to $n + 1$ terms. For example, $3x = 0$, with 1 term, is an algebraic equation of degree 1. The equation $x^4 - 2 = 0$ with 2 terms has degree 4. In fact, the general algebraic equation of degree n can be expressed as the product of n factors, each displaying one of its n roots, $r_1, r_2...r_n$. Type equation here.

$$a_0 (x - r_1)(x - r_2)\cdots(x - r_n) = 0$$

It is important to note that not all the roots are real numbers, nor are they all different. From the algebraic equations $x - n = 0$, $n = 1, 2, 3...$, we can see that the solutions $x = 1, 2, 3...$ form the set of natural numbers. It is clear that the set of natural numbers is a proper subset of A. From the equation $qx - p = 0$ or $x = p/q$, where p and q are integers and q is not zero, we see that the set of algebraic numbers (A) contains the rational numbers (F) as a subset. From the equation $x^2 - 2 = 0$, with root $x = \sqrt{2}$, and the fact that $\sqrt{2}$ cannot be expressed as a rational p/q (see appendix 4), we can state that the rational numbers are a proper subset of A. The question arises, Does the set of algebraic numbers (A) have a cardinal number larger than N_0?

If we use the common measure for size—i.e., A is larger than B if A has more elements than B—we can say that the set of algebraic numbers (A) is larger than both the rational and natural numbers. However, we are dealing with transfinite cardinal numbers, where the normal measures for size serve only as an intuitive guide. We have shown that the set of natural numbers and the set of rational numbers have cardinal numbers N_0. We know that there is an infinite number of equations yielding algebraic numbers as roots. Each equation of degree n has n roots, and depending on the values of the a_i, each equation of degree n may contain from 1 to $n + 1$ term. For example, $x^4 + x^2 + 1 = 0$ has three terms and $x^4 + 2 = 0$ has 2 terms. Both are equations of degree 4. Each equation contributes 4 roots (possible members) to the set (A). Yet with all the possible forms of equations with degree n—$n = 1, 2, 3, 4...$—contributing elements to A, we can still show that the set (A) has cardinal number N_0.

To do so, we followed the procedure used to demonstrate that the rational numbers have the same cardinal number as the natural numbers do. With the rational numbers, we were able to divide the rational numbers into finite sets, order the numbers in each set, and then order the sets. We follow a similar procedure to distinguish one algebraic equation from another. The distinguishing features of algebraic equations are their degree (n) and coefficients (a_i), where $i = 0, 1, 2, 3,...n$. Using the index

$$|a_0| + |a_1| + |a_2| + |a_3| +...+ |a_{n-1}| + |a_n| + n,$$

we can segregate the universe of algebraic equations into subsets of equations having the same index. We can then order the solutions to the equations in each subset and then place all the solutions in a natural order. Since $a_0 > 0$ and n has to be at least 1, there are no equations having index 1. The only equation of index 2 is $x = 0$, where $a_0 = 1$ and $n = 1$. The sets of equations for indices 2, 3, 4, and 5 follow. We determine the roots for each of the equations, as shown below:

$x = 0$ has root 0   Index 2

$x + 1 = 0$, $x - 1 = 0$ has roots 1, −1   Index 3
$x^2 = 0$ has root 0

$x^2 \pm 1 = 0$ has roots $\sqrt{(-1)}, \sqrt{(-1)}, 1, -1$ Index 4
$2x \ 1 = 0$ has roots $\frac{1}{2}, -\frac{1}{2}$
$x \pm 2 = 0$ has roots $2, -2$
$x^3 = 0$ has roots $0, 0, 0$

$x^3 \pm 1 = 0$,   Index 5
$x^2 \pm 2 = 0$ has roots $\sqrt{2}, \sqrt{-2}$
$2x^2 \pm 1 = 0$ has roots $\sqrt{2}/2, \sqrt{-2}/2$
$3x \ 1 = 0$ has roots $\frac{1}{3}, -\frac{1}{3}$
$x \pm 3 = 0$ has roots $3, -3$
$x^4 = 0$

We start with index 2 and order the roots according to their index. The nonreal roots and those that were listed earlier in the compilation are discarded. A typical order follows:

0, 1, 1, 1/2, 1/2 , 2, 2, 1/3, 1/3, √2/2, √2/2, √2, √2, 3, 3, ...

From the list of roots, we form a sequence (A) of roots wherein every root has an index and a position in that index set of roots. Thus, starting from index 2, we can label the roots in a definite order ($r_1, r_2, r_3...$). We obtain the one-to-one correspondence when we pair each root with its subscript (r_i, i), where $r_i \in A$ and $i \in N$.

NUMBERS

APPENDIX 4

Square Root of 2

WE HAVE STATED that the square root of 2 is not a rational number; i.e., it cannot be expressed as the ratio of two integers, p/q, where q is not 0. In geometry, we learned that two line segments are commensurable if it is possible to find a shorter line segment that can be paced off exactly a natural number of times over the length of the two line segments. For example, two line segments of length (3 and 4 inches) are commensurable since there are line segments of shorter length (i.e., ½) that can be used to pace off (exactly cover) both line segments. If we consider the line segments of length, 3 and 4, as the sides of a rectangle, its diagonal $5 = \sqrt{(3^2 + 4^2)}$ is commensurable with both its sides. Is it possible to construct other rectangles whose diagonals are commensurable with their sides? The answer is yes. There exist formulas for determining three numbers that serve as the commensurable sides of a rectangle and its diagonal. However, the diagonal of a square in never commensurable with its side. This fact is very simple to prove. Consider a square with sides equal to x and diagonal y. Then

$$y^2 = x^2 + x^2, \text{ or } y^2 = 2x^2, \text{ or } y = \pm\sqrt{2}\, x.$$

This states that the length of a diagonal of a square is always equal to the $\sqrt{2}$ times the length of the side. Hence, we need show that the $\sqrt{2}$ is irrational. To show that the square root of 2 is not a rational natural number, let us assume that it is; i.e., we can write $\sqrt{2} = p/q$, where q is not 0 and p and q have been reduced to their lowest form (i.e., they have no common factor). Squaring both sides, $p^2 = 2q^2$. Thus, p^2 is an even number since it has the factor 2. We also conclude that p is even. For if it were odd, we could write it as $p = 2r + 1$, where r is a positive integer. On squaring, $p^2 = 4r^2 + 4r + 1$ yields an odd number, contradicting that

p^2 was even. With p even, we can write it as $2y$, where y is an integer. Substituting $2y$ for p, we can write $(2y)^2 = 2q^2$ or $q^2 = 2y^2$. Using the same argument as above, we conclude that q is even. We have shown that p and q are both even. Hence, they have a common factor 2, contradicting our original premise.

APPENDIX 5

Irrational and Transcendental Numbers

THE SET OF irrational numbers (I) is defined by the set equation $I = R - F$, that is, the set of numbers that remain when the rational numbers are removed from the real numbers. The set of transcendental numbers (T) is defined by the set equation $T = R - A$, that is, the set of numbers that remain when the algebraic numbers are removed from the real numbers. In the sequel, we show how the Cantor diagonal process is applied repeatedly to the set of rational numbers (F) to construct a set of irrational numbers (I). The process is depicted graphically in the following figure, A5.1.

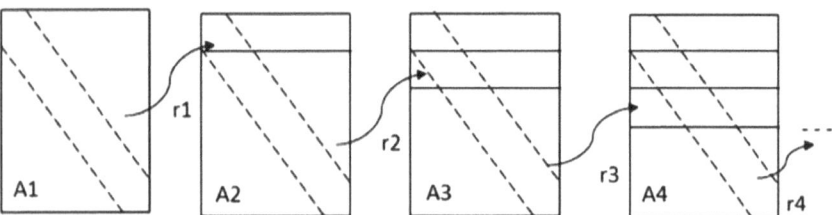

Figure A5.1. Cantor's diagonal process

0.4 9 9 9 9...	0.1 1 1 1 1...	0.2 1 1 1 1...	0.1 2 1 1 1...
0.3 3 3 3 3...	0.4 9 9 9 9...	0.1 1 1 1 1...	0.2 1 1 1 1...
0.2 4 9 9 9...	0.3 3 3 3 3...	0.4 9 9 9 9...	0.1 1 1 1 1...
0.6 6 6 6 6...	0.2 4 9 9 9...	0.3 3 3 3 3...	0.4 9 9 9 9...
0.1 9 9 9 9...	0.6 6 6 6 6...	0.2 9 9 9 9...	0.3 3 3 3 3...
....	0.1 9 9 9 9...	0.6 6 6 6 6...	0.2 9 9 9 9...
....	....	0.1 9 9 9 9...	0.6 6 6 6 6...
A 1...............	A 2...............	A 3...............	A 4...............

In appendix 2, we illustrated how to arrange all fractions x/y as solutions to $x + y = n$, $n = 1, 2, 3\ldots$ in a sequence

$$0, 1, \tfrac{1}{2}, 2, \tfrac{1}{3}, 3, \tfrac{1}{4}, \tfrac{2}{3}, \tfrac{3}{2}, 4, \tfrac{1}{5}\ldots$$

From the sequence, we select, in order, the fractions with values between 0 and 1, yielding

$$\tfrac{1}{2}, \tfrac{1}{3}, \tfrac{1}{4}, \tfrac{2}{3}, \tfrac{1}{5} \ldots$$

We convert the fractions to their decimal forms and list the decimals into an array as follows:

$$0.49999\ldots$$
$$0.33333\ldots$$
$$0.24999\ldots$$
$$0.66666\ldots$$
$$0.19999\ldots$$
$$\ldots\ldots\ldots$$

Let us call this array A_1. Apply the Cantor diagonal process to the array A_1. Form a new number using the digits on the diagonal of the array. The result is 43969. The diagonal numbers are shown in bold in the matrix of numbers. We modify the digits of the diagonal number as follows: if the digit is a 1, change it to a 2; if it is not a 1, change it to a 1. The result is the number 11111…. Label this number r_1. Place r_1 at the head of array A_1, resulting in a new array, A_2.

$$0.11111\ldots$$
$$0.49999\ldots$$
$$0.33333\ldots$$
$$0.24999\ldots$$
$$0.66666\ldots$$
$$0.19999\ldots$$
$$\ldots\ldots\ldots$$

Again, apply the Cantor diagonal process to A_2 and obtain $r_2 = 0.21111\ldots$

Placing $r_2 = 0.21111...$ at the top of A_2, we obtain the array A_3.

$$0.21111...$$
$$0.11111...$$
$$0.49999...$$
$$0.33333...$$
$$0.29999...$$
$$............$$

As shown in the figure A5.1, this process of forming numbers r_i, $i = 1, 2, 3,...$ and arrays A_{i+1} $i = 1, 2, 3...$ can be continued indefinitely. As formed, the r_i are all different from one another. They differ from each element of F in at least one digit. Thus, the r_i are not elements of F and are therefore irrational numbers.

In the set of real numbers, the transcendental numbers are real numbers that are not algebraic. The Naperian base e and pi are transcendental numbers. The square root of 2 is irrational but not transcendental. While the set of transcendental numbers has cardinal number c, the same as the real numbers, very few are known. To show that any given number is transcendental can be a very difficult process.

APPENDIX 6

Line and Plane Have Same Cardinal Number

YOU WOULD THINK that the set of points in a two-dimensional unit square, which is shown below, would have a larger cardinal number than the set of points in a one-dimensional line segment of unit length. With the real numbers on the unit length line segment having cardinal number c, you would expect, keeping with the old but erroneous sizing of infinities, the two-dimensional plane to have cardinal number c^2. Again, our intuition fails us. The set of points in the unit square has the same cardinal number c as the unit line segment of real numbers.

Figure A6.1. Line and square

To show that the set of points in the two-dimensional square has the same cardinal number as the set of points in the line segment, we will restrict our application of the one-to-one correspondence to a unit square, as shown in the figure below. The correspondence applied to the unit square and line segment applies equally well over the entire plane and the x-axis. The set of points in the unit square is defined by $P = \{(x, y) / 0 < x < 1, 0 < y < 1\}$. The set of points in the unit line segment of real numbers is defined by $L = \{x / 0 < x < 1\}$. The decimal form of each real number in L is given by

$$x = 0.a_1\, a_2\, a_3\, a_5\, a_6\, a_7\, \ldots.$$

We start with the decimal form for a real number on the unit line segment and show how to partition the digits of the real number into smaller groups of digits. After the partitioning into groups, we use the groups to construct the coordinates for a point in the unit square. We illustrate the partitioning and assignment process using the following example:

Let $r = 0.5\ 1\ 0\ 3\ 6\ 0\ 7\ 1\ 2\ 5\ \ldots$

Starting with the most significant digit, partition off each digit that is not 0. In r, 5 and 1 would be the first two groups. When you reach a 0, continue until you reach a nonzero. Partition off that sequence ending in a nonzero digit. In r, 03 would be the third group. Underline each partitioned group, and assign numbers to the groups as shown in figure A6.2.

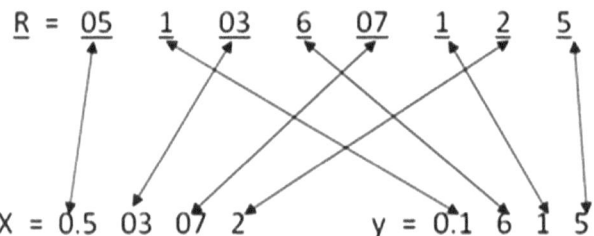

Figure A6.2. Line and plane have same cardinal number.

We form the coordinates of a point in the unit square by matching the groups assigned odd numbers to the point's *x*-coordinate and groups assigned even numbers to the *y*-coordinate. We have drawn lines to show the one-to-one correspondence. It should be clear that we could have reversed the one-to-one correspondence process by first partitioning the decimal forms for the coordinates of the point in the square and then applying the groups to form the decimal for a point in the line segment.

Through the above process, we have shown that a one-dimensional line in the plane has the same cardinal number as the two-dimensional plane.

INDEX

A

aleph null, 76, 103
algebraic equations, 116
algorithms, 24, 86, 96
Analysis
  deductive, 17-18
  descriptive, 16-19
  normative, 16-17
  predictive, 16-18, 20
  prescriptive, 16-17, 19-20
arguments
  deductive, 59-60
  valid, 47-48
axioms, 71
axiom systems, 75

B

bar, 103
base 2 system, 89
base 10 system, 89-90
binary digits, 92
Binary multiplication, *94*
binary numbers, 92, 94-95, 97

C

Cantor, Georg, 82
  diagonal process, 89
cardinal numbers, 29, 102-3
complement of sets, 35, 38, 72, 89, 103

condition, 47
  necessary, 46-47, 52, 67
  sufficient, 46-47, 52, 55, 67
conditional statement. *See*
  implication
connectives, 38-39, 41, 43, 46
contrapositive, 45-46, 82
conversion, 94-95, 99
  binary to decimal, 94
  decimal to binary, 101
conversion processes, 96, 99
credibility, 58
  code, 50-51, 54
  patterns, 61
  relationships, 63
  scale, 58

D

decimal, 92
  cyclical, 109
  number, 71-72, 77, 82, 89, 92, 99
  rational, 99
deductive patterns, 65-66

E

element, 29-39, 47, 73-77, 83,
  102-5, 116, 122
Euclid, 88
evaluation, 50-54, 57, 59-60, 67-68

H

highest power division algorithm, 96, 99

I

implication, 42-44, 46-47, 60
information credibility, 52, 57
information synthesis, 52
integers, 77
irrational numbers, 32, 72, 84, 103, 120

K

Kent chart, 54-55

L

logic, elements of, 39
logical connectives, 38, 41
logical structures, 21, 39, 41

N

Naperian base, 122
numbers
  algebraic, 32, 72, 79, 81
  binary, 92, 94-95, 97
  cardinal, 29, 102-3
  composite, 87, 89
  converting decimal, 99
  irrational, 32, 72, 84, 103, 120
  natural, 27, 30, 32, 71, 74, 77, 81
    adjacent, 79
    consecutive, 114
  prime, 32, 86-87
  rational, 32, 78, 81, 89
  real, 32, 80
  transcendental, 32, 72, 83, 103, 120, 122
number systems, 89

O

one-to-one correspondence, 28, 72

P

patterns
  credibility, 61
  cyclical, 99-100
  deductive argument, 60-62
plausible reasoning, 49-50
polynomial equation, 80
properties, 18, 29-31, 33, 39, 72-73, 86

S

scales
  credibility, 58-59
  plausibility, 54-55
sequence
  cyclical, 109
  terminating, 109
sets
  complement, 35, 38, 72, 89, 103
  definitions, 33-34
  difference, 37
  disjoint, 36
  notation, 29, 31, 33-37, 39, 42-43, 81, 84, 114
  null, 32
  theory, 30-31, 38

square root, 118
statements
  biconditional, 46
  compound, 39-42, 44-45, 47-49
  conditional, 42-43, 46
  contradictory, 44, 83, 118-19
  contrapositive, 45-46, 82
  converse, 34, 45-46
  equivalent, 44-46
  implication, 42, 44, 47, 60
  inverse, 45
  simple, 39-41, 43, 50
subset, 31-35, 37-38, 71-72, 75-77, 80-81, 88-89, 102-6, 115-16

T

transcendental numbers, 23-24, 32, 72, 83-84, 89, 102-3, 120, 122
truth tables, 41-42
  ˜A ˆ˜B, 44
  A—B, 46
  combination, 42
  contrapositive, 45
  converse, 45
  if A, then B, 42-43
  inverse, 45

V

valid arguments, 47-49, 62-63, 67
Venn diagrams
  and, 38, 43, 69
  complement, 35
  difference, 37
  disjoint, 36
  intersection, 36
  subsets of R, 81

www.ingramcontent.com/pod-product-compliance
Lightning Source LLC
Chambersburg PA
CBHW030816180526
45163CB00003B/1302